ORCHIDS
OF
SAMOA

ORCHIDS
— OF —
SAMOA

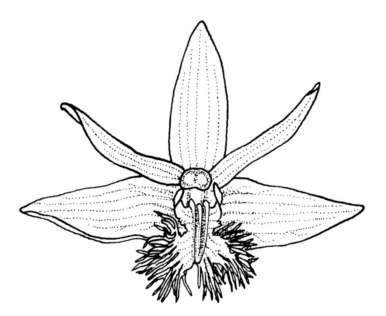

Flickingeria comata

by Phillip Cribb and W. Arthur Whistler

Illustrations by Sue Wickison
and Susanna Stuart-Smith

Royal Botanic Gardens, Kew

First published 1996

Addresses of Authors:

Phillip J. Cribb, The Herbarium, Royal Botanic Gardens, Kew, Richmond, Surrey TW9 3AE, U.K.

Arthur W. Whistler, Botany Department, University of Hawai'i, 3190 Maile Way, Honolulu, Hawai'i 96822, U.S.A.

Cover design and page make-up by Media Resources, Information Services Department, Royal Botanic Gardens, Kew.

ISBN 1 900347 01 6

Printed and bound in Great Britain by
Whitstable Litho Printers Ltd

ACKNOWLEDGEMENTS

We would like to thank Mark Clements, Wolfgang Gerlach, Paul Kores, Paul Ormerod, Sarah Thomas and Jeffrey Wood for their help on taxonomic matters; G. Hermon Slade and Gerald McCraith who stirred the interest of one of the authors (PC); Alec Pridgeon for carefully editing the text; Mike Lock for assistance with the preparation of the colour plates and for editorial advice; Milan Svanderlik and his team for help with the maps; Sarah Thomas for compiling the index; Sue Wickison, Susanna Stuart-Smith and Mair Swann for the line drawings.

We are also grateful to the Curators of the following herbaria for access to or the loan of specimens: AMES, BISH, BM, G, HAW, HBG, K, P, UC, W, Z.

We are particularly grateful to Paul Ormerod for extensive comments and advice on the manuscript.

CONTENTS

INTRODUCTION

Samoa is a volcanic archipelago situated in the south Pacific Ocean at a latitude of 13–15° south (11° when Swains Island is included) and a longitude of 168–173° west, and runs in a west-northwest direction east of Fiji, north of Tonga, south of Tokelau, and west of Niue and the Cook Islands (Map 1). Its nine inhabited islands and several uninhabited islets, plus two distant coral islands (Swains Island, which is inhabited, and Rose Atoll, which is not) have a total area of *c.* 3100 km².

The archipelago is divided politically into Western Samoa, which is an independent country, and American Samoa, which is an unincorporated territory of the United States; the two are separated by a strait 64 km wide (Map 2). Western Samoa is by far the larger of the two with about 93% of the land area of the archipelago. It comprises two main islands, Savai'i (1820 km² area, 1860 m elevation) (Map 3) and 'Upolu (1110 km², 1100 m) (Map 4), which are separated by a strait 21 km wide. Between the two are the small, uninhabited islands of Apolima and Manono, which have a combined area of *c.* 6 km² Four additional islets, Nu'utele, Nu'ulua, Namu'a, and Fanuatapu, known collectively as the Aleipata Islands, are situated off the eastern end of 'Upolu; these are all uninhabited tuff cones with a total area of less than 4 km².

American Samoa consists of five main volcanic islands and two atolls. The largest of these is Tutuila (124 km² area, 650 m elevation) (Map 5); lying off its southeastern end is the small tuff cone island of 'Aunu'u, which has an area of less than 2 km². Approximately 100 km to the east lies the group of islands known as Manu'a (Map 6) which comprises Ta'u (39 km², 960 m), Ofu (5 km², 495 m), and Olosega (4 km², 640 m). Approximately 140 km to the east of Ta'u lies the uninhabited Rose Atoll, and 320 km north lies Swains Island, which is home to a small population. Swains, which is politically a part of American Samoa, geographically belongs with Tokelau. The current population of Western Samoa is over 170,000, and of American Samoa, over 46,000.

Physical Geography

Samoa is volcanic in origin and "oceanic", that is, it was formed from basalt rising from the ocean floor of the Pacific basin beyond the continental islands to the west. The archipelago was born in isolation and has never had a connection to any other land area. The islands originated in the Pliocene (5.2–1.6 million years B.P.), and were formed generally in a westerly direction, with the youngest islands on the west end of the chain (Savai'i) and the oldest on the east (Rose Atoll).

The archipelago is still volcanically active; two eruptions have been recorded in historic times. The older of the two occurred in about 1760, and the younger one between 1902 and 1911. In addition to the lava flows, several tuff

1

Map 1. Position of Samoa in the South West Pacific

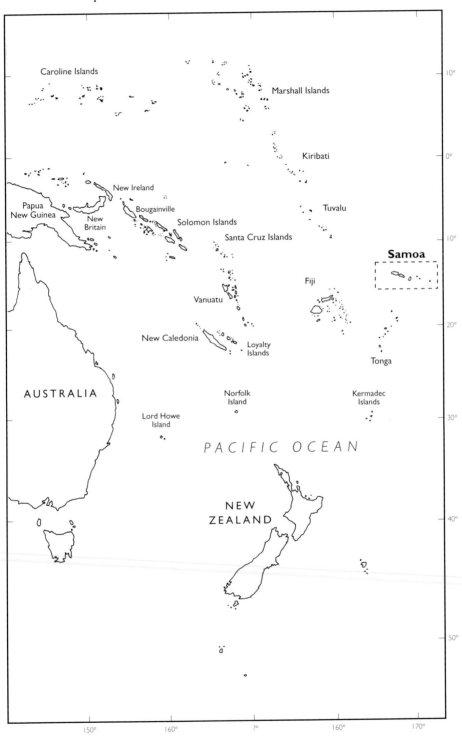

Map 2

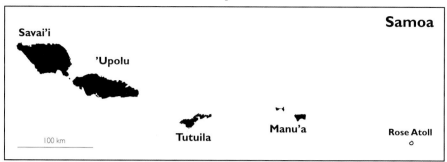

Map 3

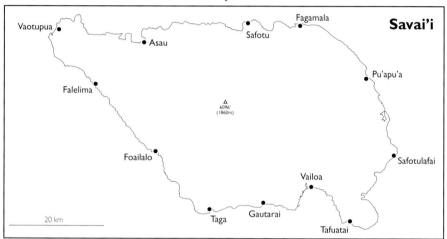

Map 4

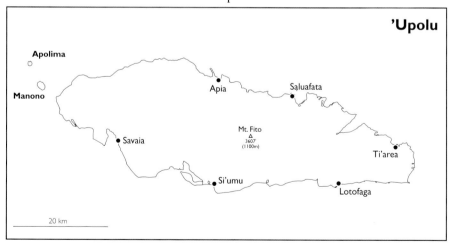

Map 5

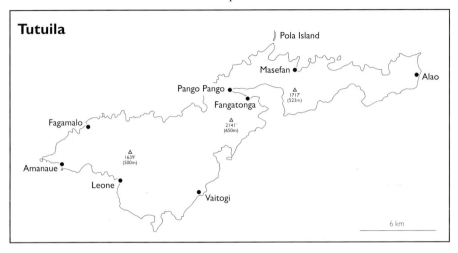

Tutuila

Pola Island

Masefan

Alao

Pango Pango

Fangatonga

1717'
(523m)

Fagamalo

2141'
(650m)

1639'
(500m)

Amanaue

Leone

Vaitogi

6 km

Map 6

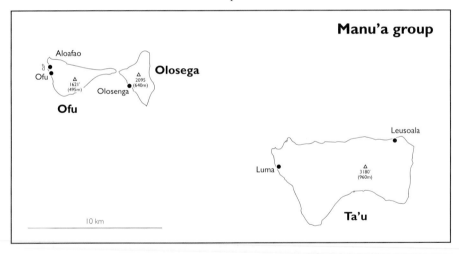

Manu'a group

Aloafao

Ofu

Olosega

1621'
(495m)

2095
(640m)

Olosenga

Ofu

Leusoala

Luma

3180'
(960m)

10 km

Ta'u

cones are found in coastal areas. These were formed when upwelling lava
came into contact with saline groundwater, causing tremendous steam
explosions that deposited a mixture of limestone (from reefs) and volcanic
rocks. The best examples of tuff cones are Apolima Island, the four Aleipata
Islands, and 'Aunu'u Island. The two atolls in Samoa, Swains and Rose, are
composed of sandy deposits on a coral cap topping ancient, eroded or
submerged oceanic volcanoes.

Geologists divide the irregularly occurring periods of volcanic activity in Western Samoa into six "volcanics"; these are, from oldest to youngest, the Fagaloa, Salani, Mulifanua, Lefaga, Puapua, and A'opo volcanics, named from localities where typical examples occur (Kear and Wood 1959). The oldest rocks, those of the Fagaloa volcanics, predominate on the northeastern quarter of 'Upolu, but are rare on Savai'i. The topography of these areas is steep, and the soils are deep and weathered. The next oldest rocks, those of the Salani volcanics, predominate on the southeastern quarter of 'Upolu, and on parts of the eastern half of Savai'i. The Mulifanua volcanics predominate in the northwestern quarter of 'Upolu and the western half of Savai'i. The Lefaga volcanics are of minor importance and occur only on 'Upolu, mostly on the western half of the island. The Puapua volcanics are typical of many of the coasts of Savai'i, but on 'Upolu are restricted mostly to the south-central coast. The most recent rocks, those of the A'opo volcanics, comprise two major lava flows on the north half of Savai'i.

Geologists recognize a different series of volcanics on Tutuila (Stearns 1944) and Manu'a (Stice and McCoy 1968), but these are not correlated with those of Western Samoa, and are necessarily much smaller in extent.

Climate

Since Samoa is situated between the tropic of Capricorn and the Equator, its climate is tropical. Typical day temperatures (at sea level) are between 24–29°C. The difference between the average temperature of winter (June to September) and summer (December to March) is only about 2°C, and the average annual temperature is about 26°C (Wright 1963). The relative humidity is constantly high, averaging about 80%. Rainfall is heavy throughout the archipelago, with a minimum of 200 cm in all places on the main islands. The Western Samoa capital of Apia, for example, which is situated at the middle of the north coast of 'Upolu, receives about 290 cm of precipitation annually. Half of the rainfall in Samoa occurs from December to March, but there is no real dry season, since even in the driest places (which are on the leeward or north and west sides of the islands) all months average over 10 cm of precipitation. Droughts of varying duration occasionally occur, but these do not have much permanent effect on most of the vegetation. Periodic hurricanes hit the islands; although there has been a recent rash of destructive hurricanes (Tusi 1987, Ofa in 1990, and Val in 1991), serious ones average less than one a decade.

Vegetation of the Islands

Samoa was originally covered almost entirely with rainforest; the only types not classified as such (Whistler 1992a) were littoral communities (herbaceous strand, littoral shrubland, *Pandanus* scrub, and littoral forest), wetlands (freshwater swamps, mangrove swamps, coastal marshes, montane marshes, and montane bogs), volcanic scrub (lowland and upland) upland scrub vegetation (montane scrub, summit scrub), and secondary vegetation (caused by hurricanes, etc.).

The human disturbance over the 3000 years of human occupation has led to the loss of much of the native vegetation. Compared to other Polynesian islands, Samoa, with a large area and high elevation, is somewhat less disturbed, although over half of its native vegetation has been severely altered by human activity and natural catastrophes. Much of this has happened in the last few decades, as an increasing population, a major forestry industry, and three destructive hurricanes have taken their toll.

With a maximum elevation of over 1800 m, a wide ecological variation exists in Samoa that allows for the profusion of native plant species. The rainforest is not homogeneous, but can be divided somewhat loosely into several communities or associations distinguished by differences in flora, elevation, topography, and soil.

The island of Savai'i, the largest in the archipelago, is volcanically recent and the soils quite rocky. Nearly the entire lowlands (except for the drier northwest quarter) were once covered with *Pometia pinnata* (Sapindaceae) lowland forest, but most of this has been removed by recent logging. The highlands of the island comprise the best remaining undisturbed vegetation in Polynesia, but recent hurricanes have caused much damage, and the lower portions have been recently logged.

'Upolu, the second largest island of the archipelago, consists mostly of fertile, gently sloping lowlands that have now mostly been converted to plantations. The remaining lowland forest on the island is mostly on the ridges on the northeast corner of the island, which are dominated by *Calophyllum neo-ebudicum* (Clusiaceae), *Canarium vitiense* (Burseraceae), *Syzygium inophylloides* (Myrtaceae), and *Intsia bijuga* (Leguminosae), and the geological recent lava flow on the south-central coast, which is covered by *Pometia pinnata* (but was devastated by the recent hurricanes). The interior of the island rises to over 1100 m elevation where montane and cloud forest prevail. These communities contain numerous plant species, including most of the native orchids.

Tutuila, which is the most rugged island in the archipelago, is largely disturbed, but native forest can be found along the north-central coast of the island. This area comprises ridges dominated by a combination of lowland forest trees similar to that found on the northeast 'Upolu ridges. *Pometia pinnata* lowland forest formerly existed on a recent volcanic plain, but this has been now almost entirely removed. Relatively few orchids are found in the lowland forest, because of the low elevation and moderate rainfall, but the montane scrub community which lies atop several "trachyte plugs" scattered throughout the island, is made up of a unique scrubby vegetation ideally suited to the growth of epiphytic orchids.

Ta'u, the smallest of the four main islands, extends up to 960 m; a unique "summit scrub" community dominated by climbers and shrubs covers the upper portion (Whistler 1992b). Because of the steep topography and relatively small (and currently decreasing) population, much of the island is covered with native vegetation, although this was devastated by the recent hurricanes. About half of the area of the island is a proposed National Park. At least 44 orchids are known to occur on the island.

The Origins of the Vascular Flora

The native flora of Samoa consists of about 230 ferns and "fern allies", and 540 species of flowering plants. Samoa has the second largest flora in tropical Polynesia (behind Hawai'i), but it is only about one third as rich as that of Fiji located 700 miles to the west. Two thirds of the flowering plant species are dicotyledons and one third are monocotyledons and these fall into 95 plant families and about 300 genera. The level of endemism of the flowering plants is estimated to be about 30% (Whistler 1992a) at the species level, but only a single genus, *Sarcopygme* (Rubiaceae), is endemic to Samoa. Another 250 or so intentionally or accidentally introduced plants have become naturalized, most of which are usually referred to as "weeds" (Whistler 1988). The plant families with the highest number of native species are Orchidaceae (101), Rubiaceae (*c.* 46), Urticaceae (*c.* 25), Leguminosae (*c.* 22), Myrtaceae (*c.* 21), Euphorbiaceae (*c.* 21), and Gramineae (*c.* 21 including Polynesian but not recent introductions). Four epiphytic orchids common in mangrove and lowland forests are shown in Fig. 5.

Since Samoa is a relatively young oceanic archipelago, there has not been sufficient time for much speciation to occur. The exceptions are the genera *Cyrtandra* (19 of the 20 species are endemic), *Psychotria* (17 of 20), and *Elatostema* (14 of 14). Other large genera with lower rates of endemism include *Syzygium* (16 species), *Bulbophyllum* (11), *Dendrobium* (14), and *Ficus* (8). Most of the flora is comprised of recent arrivals, mostly from the west. Nearly two-thirds of the native species of Samoa are found in Fiji (Smith 1979–1981), as are the vast majority of the genera and the orchids.

Botanical Exploration of Samoa

The collection of the Samoan flora began dramatically when a botanist from the La Pérouse expedition gathered a number of specimens in the vicinity of A'asu Bay on the north coast of Tutuila in 1787. However, his specimens, which he carried on his back when he swam to the ship during a battle between Samoans and his French shipmates, were lost several months later along with the ship and crew. The next botanical collection was not until 1838, and it was also made by a French Expedition, this one under the command of Dumont d'Urville. The following year the United States

Exploring Expedition (U.S.E.E.), under the command of Capt. Charles Wilkes (1798–1877), visited Samoa, and a large collection was made (Pickering 1876).

Following this exploratory phase of Samoan botany, no collecting was done for the next two decades until a second phase began, one characterized by the work of amateur European botanists. The most notable of these are Eduard Graeffe, a Swiss physician (who worked in Samoa from in *c.* 1862–1872), the Rev. Thomas Powell, an English missionary (1809–1887), and the Rev. Samuel Whitmee (1838–1925), another English missionary. Graeffe and Whitmee did not publish their botanical information, and Powell published only a list of Samoan plant names.

A third phase of Samoan botany dates from about 1893 to 1906, and is characterized by extensive collections made by German and Austrian botanists. The first of these was Franz Reinecke (1866–?), who collected in Samoa from 1893 to 1895, and published the first flora of Samoa, *Die Flora der Samoa-Inseln* (1896, 1898). He was followed by Friederick Vaupel (1876–1927), a German physician and amateur botanist whose collections date from 1904 to 1906, and Karl Rechinger who visited Samoa in 1905 with his wife, Lily, and published a series of reports (1907–1915) based on their collections.

The last and current phase of Samoan botany began in 1920 with the work of W.A. Setchell (1924) in American Samoa, and is characterized by collections made by amateur and professional American or American-sponsored botanists. Erling Christophersen, who collected in Samoa in 1929 and 1931, published *Flowering plants of Samoa* (Christophersen 1935, 1938), the most comprehensive and meticulous flora. Shortly after that, Truman Yuncker collected in American Samoa, and published *Plants of the Manua Islands* (1945). Since Yuncker's time, several significant collections have been made, most notably by M. Bristol in 1968, P.A. Cox in the 1980s, and W.A. Whistler from 1971 to the present, but little has been published about these, other than checklists of the flora of the Aleipata Islands (Whistler 1984b), Tutuila (Whistler 1994), Ta'u (Whistler 1992b), and American Samoa (1980), and a partial checklist of the montane region of Savai'i (Whistler 1978).

History of Samoan Orchids

The first collections of orchids from the Samoan archipelago were made by Charles Wilkes who led the U.S. Exploring Expedition to the Pacific Islands between 1838 and 1842 (Pickering 1876). He was followed there by Eduard Graeffe, a Swiss physician who spent 11 years in Samoa working for the firm of Johann Cesar Godeffroy of Hamburg. H.G. Reichenbach, the German orchid specialist, described three of Graeffe's orchids from Samoa as new in his account of the orchids of the neighbouring Fiji Islands in 1868. Reichenbach was the most active orchid taxonomist of the day and most orchid novelties were sent to him until his death in 1885. He described a further seven

Samoan orchids based on collections by Wilkes, Graeffe, Powell and Whitmee including *Calanthe alta, Phaius graeffei* (= *P. terrestre*), *Platylepis heteromorpha* (= *Moerenhoutia heteromorpha*) and *Malaxis heliophila* (= *Oberonia heliophila*) (Reichenbach 1877, 1878). Powell stayed in Samoa for 25 years from 1860 until 1885 while Whitmee had two spells there from 1863 until 1877 and from 1891 until 1894.

Baron Ferdinand von Mueller, the distinguished German botanist who became Director of the Melbourne Botanic Garden, described five Samoan orchids, three of which were new (Mueller 1881). These had been collected in 1880 by his fellow countryman Ernst Betche (1851–1913). Betche later settled in Sydney where he remained for the rest of his life. Two further species collected by Betche were described by Mueller and Fritz Kraenzlin (1893, 1894).

Kraenzlin produced the first attempt at listing the Samoan orchid flora (Reinecke 1898). He recognised 52 species, seven new to science, in 25 genera. The novelties and many of the additions were based on the collections of Franz Reinecke who was in Samoa from 1893 until 1895. Kraenzlin compared these with Malaysian species but time has reflected badly on his taxonomy and many of his determinations are now considered erroneous. A further species, based on a collection by a Mr. Walter, was added by Robert Rolfe, the Kew orchid specialist, in 1898.

The new century saw a renewed interest by German botanists in Samoan orchids. Rudolf Schlechter (1872–1925), the most prolific orchid taxonomist of the time, described three novelties in 1906 based on collections by Betche and Whitmee. However, the most significant collections of Samoan orchids were made between May and August 1905 by the Austrian botanist Karl Rechinger and his wife Lily. The results were published by Hans Fleischmann and Rechinger (1910). Forty-one species in six genera were identified and one new genus, *Coralliokyphos,* and six new species described. Many of their determinations are now considered incorrect but the extent of Rechinger's collections allowed Schlechter in 1910 to undertake a detailed revision of Samoan orchids that was a marked improvement on earlier attempts. He recognised 82 species in 39 genera, 31 of the species being newly described in the account and based on the collections of Rechinger, Betche and Friederich Vaupel (1876–1927). Schlechter added a further novelty *Coelogyne whitmeei* (= *C. lycastoides*) to the flora in 1912.

The bombing of the Berlin-Dahlem Herbarium during the second World World was a major catastrophe that interrupted the study of orchid taxonomy because of the total loss of the orchid herbarium collections, notably the immense number of types described by Rudolph Schlechter and those of Kraenzlin. The Samoan orchids were amongst these. Confirmation of the identity of some of Schlechter's and also Kraenzlin's species is a continuing problem for all orchid taxonomists working in the region. Nevertheless, there has been some progress in the study of Pacific Island orchids in recent years as more collections have been made and more taxonomic studies carried out (Hallé 1977; Lewis & Cribb 1989, 1991; Kores 1989, 1991).

The major collections since Schlechter's account of 1911 have been made by the American-based botanists Erling Christophersen, T.G. Yuncker and by one of the authors, W. Arthur Whistler. Christophersen (1935) included 55 species in 25 genera in his account of the flora of Samoa. These were identified by Rudolf Mansfeld (1901–1960), Schlechter's successor in Berlin. He included in the account orchid collections by D.W. Garber from Tutuila and by E.H. Bryan from Tutuila and the Manu'a Islands.

Whistler's collections have added considerably to our knowledge of the orchids of Samoa, especially from the more inaccessible parts of the archipelago. Two new species, *Dendrobium whistleri* and *Bulbophyllum distichobulbum*, have recently been described by Cribb based on his collections. Whistler has kept a database of Samoan orchids for many years, updating it regularly as new collections have been made or old ones rediscovered. He has published a number of contributions to the orchid flora (Whistler 1978, 1979, 1980, 1992b, 1994).

A recent popular account of the orchids of Western Samoa by Gerlach (1994) is particularly useful for its colour photographs of 14 native species. He has also published an account of flowering behaviour of some ephemeral flowering *Dendrobium*, *Flickingeria* and *Diplocaulobium* species of Samoa (Gerlach 1992 a & b).

The Orchid Flora

The Samoan flora ranks as one of the best explored and described of those of the Pacific archipelagoes. The early work of German and British collectors and botanists and the more recent collecting by American botanists have provided a sound basis for this treatment of the orchid flora. The present account includes 101 species in 47 genera, an increase of about 20% over the previous treatment of Samoan orchids by Schlechter (1911). Bearing in mind the size of Samoa, it has a rich orchid flora. The large islands of Savai'i and 'Upolu have the largest number of species with the numbers dropping significantly on the smaller ones (see Table 1). We suspect that few additional species await discovery; perhaps another half a dozen on further exploration of some of the remoter regions.

Affinities of the Orchids

The orchid flora has close affinities with those of the adjacent larger archipelagoes of Fiji, Samoa, the Society Islands, the Solomon Islands and Vanuatu as well as with the impoverished orchid floras of the many smaller islands of the South West Pacific. Sixty-nine species are shared with Fiji; 60 with Vanuatu; 48 with the Solomon Islands; 42 with New Caledonia; 36 with New Guinea; and lesser numbers with some of the smaller archipelagoes such as

Tonga and the Society Islands (see Table 2). The affinities of the orchid flora of the region as a whole lie with that of the large island of New Guinea, some 4000 km to the west of Samoa. Samoa shares 36 species with New Guinea, while most of those Samoan orchids that are not found in New Guinea have close relatives on that island.

Table 1. The Samoan islands and their orchid flora.

Island	Size (km²)	Elev. (m)	Number of orchid species
Savai'i	1820	1860	85
'Upolu	1110	1100	78
Tutuila	124	650	48
Ta'u	39	960	43
Ofu-Olosega	9	640	26

Table 2. Affinities of the Samoan orchid flora.

Country	No. of orchid species shared	Approx. distance from Samoa (km)
New Guinea	37	4000
Solomon Islands	49	2800
Vanuatu	60	2000
New Caledonia	42	2370
Fiji	69	830
Tonga	21	570
Society Islands	9	2080
Endemic	15	–

Endemism

Considering its isolation, Samoa has relatively few endemic orchids (Table 2) and a far lower percentage than large (and much more ancient) islands such as New Caledonia and New Guinea but about the same as the smaller archipelagoes of the region. About 15% of the orchids are endemic; this may well be an overestimate because of the uncertain taxonomy and debatable synonymies of the orchids of neighbouring archipelagoes. Those species that are endemic such as *Habenaria samoensis, Liparis phyllocardia, Bulbophyllum distichobulbum, Taeniophyllum whistleri* and *T. savaiiense*, all have close allies in neighbouring islands and are best considered as neo-endemics.

Of the 46 genera only *Earina* is endemic in the South West Pacific and does not reach New Guinea. It is the only element of the peculiar orchid flora of New Caledonia that reaches Samoa.

Origins of the Orchid Flora

Apart from the recent human introductions of *Vanilla planifolia* (Fleischmann & Rechinger 1910), probably from French Polynesia where it is commercially cultivated, *Arundina graminifolia* (Gerlach 1994) from Malaya, and possibly of *Phaius tankervilleae* from South East Asia, it is probable that orchids arrived in Samoa as wind-blown seed from adjacent archipelagoes. The most likely scenario is one of the orchids island-hopping through the archipelagoes from New Guinea. The high number of species that Samoa has in common with Fiji supports this. The establishment of plants must have relied upon the incoming seed meeting a fungus that it could utilize mycotrophically. Many orchids are able to spread vegetatively by means of tubers or rhizomes. This would remove the barrier to establishment that might have arisen if suitable pollinators were unavailable in the area.

However, many Pacific Island orchids appear to be self-pollinating, at least to some degree. Examination of specimens shows that most flowers set fruit in many Samoan orchids, e.g. *Liparis condylobulbon, L. caespitosa, Oberonia* spp., *Dendrobium whistleri, Calanthe triplicata* and *Spathoglottis plicata*. We suspect that self-pollination is common in Samoan orchids and that cross-pollination may be rather a rare occurrence.

References

Christophersen, E. (1935). Flowering plants of Samoa. Family 19. Orchidaceae. *Bernice P. Bishop Mus. Bull.* **128**: 60–70.

Fleischmann, H. & K. Rechinger (1910). Botanische und Zoologische Ergebnisse einer wissenschaftlichen Forschungsreise nach den Samoan-Insel. Orchidaceae. *Denkschr. Kaiserl. Akad. Wiss. Wien, Math.-Naturwiss. Kl.* **85**: 175–432.

Gerlach, W.W.P. (1992a). Flowering behaviour of ephemeral orchids of Western Samoa 1. Flowering periodicity. *Gartenbauwissenschaft* **57(5)**: 219–222.

————— (1992b). Flowering behaviour of ephemeral orchids of Western Samoa II. Mechanism of anthesis induction. *Gartenbauwissenschaft* **57(6)**: 288–191.

————— (1994). Die Orchideenflora von West Samoa. *Orchidee (Hamburg)* **45(1)**: 7–16.

Hallé, N. (1977). *Flore de Nouvelle Calédonie et Dépendances* vol. 8. *Orchidacées.* Mus. Nationale d'Histoire Naturelle, Paris.

Kear, D. and B.L. Wood (1959). The geology and hydrology of Western Samoa. *New Zealand Geol. Surv. Bull.* **63**: 1–92.

Kores, P.J. (1989). A precursory study of Fijian orchids. *Allertonia* 5(1): 1–222.

Kores, P. (1991). *Flora vitiensis nova* vol. 5. *Orchidaceae.* National Tropical Botanical Garden, Hawai'i.

Kraenzlin, F. (1909). III. Orchidaceae Novae Samoenses. *Notizbl. Bot. Gart. Berlin-Dahlem* **5**: 109–111.

Lewis, B. & P. Cribb (1989). *Orchids of Vanuatu.* Royal Botanic Gardens, Kew.

————— (1991). *Orchids of the Solomon Islands and Bougainville.* Royal Botanic Gardens, Kew.

Mueller, F. von (1881). Record of some Orchideae from the Samoan Islands. *S. Sci. Rec.* **1**: 171.

Mueller, F. von & F. Kraenzlin (1893). *Habenaria samoensis* F.v. Mueller et Kraenzl. *Bot. Jahrb. Syst.* **17**: 487.

————— (1894). *Earina samoensis* F.v. Muell. et Krzl. *Oesterr. Bot. Z.* **44**: 211.

Pickering, C. (1876). *The geographical distribution of animals and plants in their wild state.* (From *USEE* 19, 2: 276–311). Naturalists' Agency, Salem, Mass.

Reichenbach, H.G. (1868). Orchideae. In B.Seeman, *Fl. Vitiensis*: 299.

————— (1877). Two new orchids from Samoa collected by the Rev. S.J. Whitmee. *J.Bot.* **15**: 132.

————— (1878). IV. Orchideae Wilkesianae ineditae. *Otia Bot. Hamburg.* **1**: 50–56.

Reinecke, F. (1896–1898). Die Flora der Samoa-Inseln. *Bot. Jahrb. Syst.* **23**: 237–368. 1896; 25: 578–708. 1898.

Schlechter, R. (1906). Neue Orchidaceen der Flora des Monsun-Gebietes. *Bull. Herb. Boissier* ser.2, **6**: 295–310.

————— (1910). Revision der Orchidaceen von Deutsch-Samoa. *Repert. Spec. Nov. Regni Veg.* **9**: 82–96, 98–112.

Schlechter, R. (1912). Orchidaceae novae et criticae. *Repert. Spec. Nov. Regni Veg.* **11**: 41.

Setchell, W.A. (1924). American Samoa. Part I. Vegetation of Tutuila Island; Part II. Ethnobotany of the Samoans; Part III. Vegetation of Rose Atoll. *Publ. Carnegie Inst. Wash.* **341**: (Dept. Marine Biol. **20**): 1–175.

Smith, A.C. (1979–1991). *Flora vitiensis nova: a new flora of Fiji.* National Trop. Bot. Garden, Lawai, Kaua'i, Hawai'i. 5 Vols.

Stearns, H.T. (1944). Geology of the Samoan Islands. *Bull. Geol. Soc. Amer.* **55**: 1279–1332.

Stice, G.D. and F.W. McCoy (1968). The geology of the Manu'a Islands, Samoa. *Pacific Sci.* **22**: 427–457.

Whistler, W.A. (1978). Vegetation of the montane region of Savai'i. *Pacific Science* **32** (1): 79–94.

————— (1979). Contributions to the orchid flora of Samoa. *Bull. Pacific Trop. Bot. Gard.* **9**: 34–38.

———— (1980). The vegetation of eastern Samoa. *Allertonia* **2(2)**: 46–190.

———— (1984). The vegetation and flora of the Aleipata Islands, Western Samoa. *Pacific Sci.* **37 (3)**: 227–249.

———— (1988). Checklist of the weed flora of Western Polynesia. *South Pacific Commission Techn. Paper* 194. Noumea, New Caledonia. 69 pp.

———— (1992a). The vegetation of Samoa and Tonga. *Pacific Sci.* **46 (2)**: 159–178.

———— (1992b). *Botanical inventory of the proposed Ta'u and Ofu units of the National Park of American Samoa.* Cooperative National Park Resources Studies Unit, U.S. Fish and Wildlife Service, Honolulu. 85 pp.

———— (1993). *Botanical inventory of the proposed Tutuila unit of the National Park of American Samoa.* Cooperative National Park Resources Studies Unit, U.S. Fish and Wildlife Service, Honolulu. Pp. 85–142.

Williams, L.O. (1939). Orchid studies X. *Bot. Mus. Leaflets Harvard Univ.* 7: 137– 148.

Wright, A.C.S. 1963. Soils and land use of Western Samoa. *Bull. New Zealand Soil Bur.* **22**: 1–189.

Yuncker, T.G. (1945). Plants of the Manua Islands. *Bernice P. Bishop Museum Bull.* **184**: 1–73.

THE STRUCTURE OF ORCHIDS

The diversity of form found in the orchid family in Samoa is amazing. The smallest orchid to be found there is the small *Phreatia minima* (Plate 13C) which is scarcely 2–3 cm tall with flowers 1–2 mm across, whereas the terrestrial *Corymborkis veratrifolia* (Plate 1A) may reach a metre or more in height with flowers several centimetres long. The *Taeniophyllum* (Plate 24A,B) species, *Stereosandra javanica* and *Didymoplexis micradenia* (Plate 4D) have no leaves, but most other orchids have green leaves of various shapes and sizes. Some orchids live on the ground while others grow perched on trees. What then unites these diverse plants into the family called the Orchidaceae? The distinctive features of orchids which separate them from other flowering plants lie primarily in their flowers.

The Flower

Orchid flowers are simple in structure and yet highly modified from the more typical monocotyledon flower as exemplified by a *Trillium* or *Lilium*, to which orchids are very distantly allied. These characteristically have their floral parts arranged in threes or multiples of three. Orchids are no exception. This can most easily be seen in the two outer whorls of the flower. Let us take the common Pacific Island orchid *Phaius tankervilleae*, which is similar in general floral structure to the majority of Samoan orchids, as an example. Its floral parts are situated at the apex of the *ovary* which itself can be seen to be tripartite in cross section. The outermost whorl of the flower is the calyx which consists of three *sepals* which are petal-like and coloured yellow with a red stripe in the middle. The two lateral sepals differ slightly from the third, called the *dorsal* or *median sepal*. In some orchids such as dendrobiums and bulbophyllums the *lateral sepals* form a more or less conical chin called a *mentum* at the base.

The corolla of *P. tankervilleae* comprises three *petals* which are brightly coloured. The two lateral petals, resembling the dorsal sepal in coloration and shape, are uppermost in the flower and differ markedly from the third petal which lies at the bottom of the flower. The third petal, called the *lip* or *labellum*, is highly modified, 3-lobed and has a short spur-like nectary at the base. The spur can be longer or more saccate in other orchids and can contain callosities (ridges or keels) that are diagnostic for some species. In some orchids the upper surface of the lip may be adorned with a callus of raised ridges or lamellae or tufts or areas of hairs or glands. The lip is an important adaptation of the orchid to facilitate cross-pollination. It can be imagined as a brightly coloured flag to attract potential and specific pollinators which are then guided towards the pollen and stigmatic surface by the form of the callus. The lip, therefore, can be supposed to act as a landing platform and the callus structure as a guidance system for the pollinator.

The central part of the orchid flower shows the greatest modifications to the basic monocotyledon pattern. Reduction in the number of floral parts and fusion of the male and female organs into a single structure have been the major evolutionary forces at work. The fused organ in the centre of the flower of *P. tankervilleae* is called the **column**. In this species, and in all Samoan orchids, a single **anther** lies at the apex of the column. The pollen in the anther is not powdery as in most plants, but is borne in eight discrete masses, called **pollinia**. The pollinia are attached to a sticky mass called the **viscidium**. In other species the pollinial number may be two, four or rarely six and these are attached to the viscidium either directly or by a stalk called a **stipe** in most epiphytic orchids and a **caudicle** in most terrestrial ones.

The **stigma** in *P. tankervilleae* is also positioned on the column in the centre of the flower, on the ventral surface. The stigma is a sticky lobed depression situated below and behind the anther, but in some terrestrial genera such as *Habenaria* and *Peristylus* the stigma is bilobed with the receptive surfaces at the apex of each lobe. In many species the pollen masses are transferred to the stigmatic surface by a modified lobe of the stigma called the rostellum. This is developed in *P. tankervilleae* as a projecting flap that catches the pollen masses as the pollinator passes beneath on its way out of the flower.

An interesting feature of the development of most orchid flowers is the phenomenon of **resupination**. In bud, the lip lies uppermost in the flower while the column lies lowermost. In species with a pendent inflorescence the lip will, therefore, naturally lie lowermost in the flower when it opens. However, this would not be the case in the many species with erect inflorescences, such as *P. tankervilleae*. Here the opening of the flower would naturally lead to the lip assuming a place at the top of the flower above the column. In most species this is not the case, and the lip is lowermost in the flower. This position is achieved by means of a twisting of the flower stalk or ovary through 180 degrees as the bud develops. This twisting is termed resupination.

The Inflorescence

Orchids carry their flowers in a variety of ways. Even within the same genus different species have different ways of presenting the flowers. Most Samoan orchids have inflorescences bearing two or more flowers, usually borne on a more or less elongate floral axis comprising a stalk called the **peduncle** and a portion bearing the flowers, the **rachis**. In *P. tankervilleae* the flowers are borne in an elongate erect **raceme** which is unbranched with the flowers arranged in a lax spiral around the rachis. In a raceme the individual flowers are attached to the floral axis by a stalk called the **pedicel**. In some species such as *Peristylus trasdescantifolius* pedicels are virtually absent and the flowers sessile on the axis; such inflorescences are termed **spicate**.

In the genus *Bulbophyllum* (Plates 19–22) we find some interesting variations on the multiflowered inflorescence. In several species the flowers are borne all facing to the same side of the rachis, this being called a secund inflorescence.

The most spectacular group, however, are those in which the rachis is so contracted that the flowers all appear to come from the top of the flower stalk in an umbel, with the inflorescence rather resembling the head of a daisy. Formerly these bulbophyllums, such as *Bulbophyllum longiflorum* (*Plate 19D*), were considered for this reason to be in the separate genus *Cirrhopetalum*.

Compound inflorescences with many flowers are uncommon in Samoan orchids, but where branching inflorescences are found they are termed **panicles**.

In many species the flowers are borne one-at-a-time either sessile or on a shorter or longer peduncle. Solitary flowers can be found in many genera such as *Bulbophyllum* and *Corybas*.

Vegetative Morphology

The vegetative features of orchids are, if anything, more variable than their floral ones. This is scarcely surprising when the variety of habitats in which orchids are found is considered. Orchids grow in almost every situation: on the permanently moist floor of the lowland tropical rain forest; in the uppermost branches of tall forest trees where heavy rainfall is followed by scorching sun for hours on end; on rocks; and in the grassy areas found on landslips and roadsides. The major adaptations seen in orchid vegetative morphology allow them to withstand adverse environmental conditions, in particular, the problems of water conservation on a daily and seasonal basis.

That tropical orchids might suffer from periodic water deficits is not immediately obvious. However, rainfall is not continuous even in the wettest habitats and in many places in the tropics the rainfall patterns are markedly seasonal. Furthermore, most tropical orchids are epiphytic or lithophytic, growing on the trunks, branches and twigs of the trees or on rocks. In these situations water run-off is rapid, and the orchids will dry quickly in the sunshine that follows the rain. Many orchids have marked adaptations of one or more organs which allow them to survive these periodic droughts. Some of these adaptations are as dramatic as those encountered in the Cactaceae. The stem can develop into a water-storage organ. This is so common in tropical orchids that the resulting structure has been given a technical name, a **pseudobulb**. In *Dendrobium* and *Eria* the pseudobulbs comprise several nodes while in *Bulbophyllum* they are of one node only. Pseudobulbs are also found in many terrestrial orchids and can grow either above the ground as in *Calanthe* or underground as in *Geodorum*.

A few terrestrial orchids, such as *Habenaria* (Plate 4B) and *Peristylus* (Plate 4A), lack pseudobulbs but have underground **tubers** which survive drought. The new growth grows from one end of the tuber in suitable conditions. In others, such as *Zeuxine* (Plate 3A,B) and *Goodyera* (Plate 1C), the stems are succulent but not swollen. The horizontal stem or rhizome creeps along the ground in the leaf litter, and erect shoots bearing the leaves are sent up periodically.

The leaf is another organ that has undergone dramatic modification in the orchids. Fleshy or leathery leaves with restricted stomata, such as those of *Dendrobium* and *Bulbophyllum* species, are common. The leaves of species growing in drier places can be terete as in *Schoenorchis* (Plate 23A) or *Luisia* (Plate 23B), while in *Taeniophyllum* (Plate 24A,B) the leaves have been reduced to scales and photosynthesis takes place in the often flattened green roots.

A few terrestrial orchids are also leafless and lack chlorophyll altogether and are termed **saprophytic**. The Samoan orchid, *Didymoplexis micradenia* (Plate 4D) is a saprophyte. Lacking chlorophyll, it cannot photosynthesize and must obtain all of its nutrition from the mycorrhizal fungus with which it is associated.

The autotrophic terrestrial species usually have much thinner textured leaves than their epiphytic cousins. In lowland forest, the perpetually moist atmosphere and lack of direct sunlight means that such leaves are not vulnerable to drought. Some of the terrestrial species of the forest floor have beautifully marked leaves. In *Goodyera*, *Anoectochilus*, *Erythrodes* (Plate 1B) and their relatives, the leaves can range from green to deep purple or black and may be reticulately veined with silver, gold or red.

The roots themselves are much modified in most epiphytic orchids. They provide both attachment to the substrate and also uptake of water and nutrients in a periodically dry environment. The roots have an actively growing tip; the older parts are covered by an envelope of dead empty cells called a **velamen**. The velamen protects the inner conductive tissue of the roots and may also aid the uptake of moisture from the atmosphere, acting almost as blotting paper for the orchid.

Life in the tropics can be inhospitable even for orchids. In those regions with a more marked seasonality conditions may be positively hostile for orchids at certain times of the year. Even tropical forests can have periods of relative drought where the orchids have to survive days or even weeks without rain. In these conditions, tropical orchids without water-storage capabilities in their stems or leaves can drop their leaves and survive on the moisture stored in their roots which are protected by their cover of velamen.

Conclusion

An appreciation of the vegetative structure of orchids can provide the orchid grower with the clues he needs to give his orchids optimal conditions for growth. If the seasonal nature of the growth that can be found in many orchids is ignored then they will perish rapidly. A knowledge of floral morphology is just as critical for naming orchids because they are classified into genera and species on the finer details of the structure of their sepals, petals, lip and column. Floral dissections provide the essential information for identification. For most species the shape of the sepals, petals and especially the lip will provide all of the information the reader needs. However, for the more critical taxa, details of the column, anther, pollinia and rostellum may be needed before accurate identification is possible.

ARTIFICIAL KEY TO THE GENERA OF SAMOAN ORCHIDS

1. Plants leafless or apparently so at flowering time · · · · · · · · · · · · · · **2**
   Plants not as above · **5**
2. Plants terrestrial · **3**
   Plants epiphytic; roots greenish, photosynthetic · · · · · · · · · · · · · · **5**
3. Plants buff-coloured, saprophytic · **4**
   Inflorescence green with greenish and white flowers; leaf appearing
   after flowering · **16. Nervilia**
4. Lip obovate-flabellate; callus of several verrucose ridges
   · **17. Didymoplexis**
   Lip ovate-cymbiform, obtuse; callus of 2 fleshy round basal glands
   · **18. Stereosandra**
5. Leaves absent · · · · · · · · · · · · · · · · · · · **45. Taeniophyllum**
   Leaves deciduous, falling before flowering; bracts with stipulate growths
   at base · **43. Microtatorchis**
6. Plants terrestrial · **7**
   Plants epiphytic or rarely lithophytic, rarely scandent and climbing · · **29**
7. Plants with fleshy underground tubers or pseudobulbs · · · · · · · · · · · **8**
   Plants lacking fleshy subterranean tubers or with pseudobulbs above
   ground · **13**
8. Leaf solitary, heart-shaped or ovate · **9**
   Leaves several, lanceolate to oblanceolate · · · · · · · · · · · · · · · · **10**
9. Flower solitary, helmet-shaped, subsessile on leaf · · · · · · · · **13. Corybas**
   Flower solitary or several, produced before the leaf and on a separate
   stem; not helmet-shaped · **16. Nervilia**
10. Inflorescence with apex recurved and facing towards ground
   · **47. Geodorum**
   Inflorescence erect · **11**
11. Inflorescence of many small flowers arranged in a spiral; lip entire,
   porrect · **11. Spiranthes**
   Inflorescence not obviously spirally arranged; lip 3-lobed, pendent · · **12**
12. Petals bifid, divided almost to base; dorsal sepal more than 5 mm long;
   stigmatic lobes clavate · **15. Habenaria**
   Petals entire; dorsal sepal less than 5 mm long; stigmatic surfaces sessile
   · **14. Peristylus**
13. Erect stems elongate, woody, growing from a woody rhizome · · · · · · **14**
   Erect stems fleshy, growing from a fleshy creeping rhizome or
   pseudobulbous from a creeping rhizome · · · · · · · · · · · · · · · · · **15**
14. Inflorescences lateral and terminal; flowers more than 2 cm long; bracts
   not noticeably distichous · **1. Corymborkis**
   Inflorescences terminal; flowers less than 1 cm long; bracts noticeably
   distichous · **2. Tropidia**

31. Leaves imbricate; flowers less than 3 mm across, in dense cylindrical spike ·· **25. Oberonia**
    Leaves spaced along stem; flowers more than 8 mm across ································ **35. Dendrobium (*D. goldfinchii* only)**
32. Plants sympodial, each growth determinate; inflorescences terminal and lateral ·· **33**
    Plants monopodial, each growth of indeterminate length; inflorescences lateral ·· **47**
33. Inflorescence capitate ··· **34. Glomera**
    Inflorescence one-flowered, racemose, fasciculate or umbellate ···· **34**
34. Pollinia 4 ··· **35**
    Pollinia 6 or 8 ··· **41**
35. Inflorescence lateral ·································· **39. Bulbophyllum**
    Inflorescence terminal or axillary from upper nodes ··········· **36**
36. Flowers lacking a mentum ··· **37**
    Flowers with a distinct column-foot and chin-like mentum formed by the column-foot and base of the lateral sepals ···················· **38**
37. Flowers produced one at a time, large, pale green with brown calli on lip; sepals more than 3 cm long; bracts deciduous ······· **22. Coelogyne**
    Flowers small, produced on short congested branches along the inflorescence axis; bracts persistent ···················· **33. Earina**
38. Leaf solitary ··· **38**
    Leaves 2 or more ····································· **35. Dendrobium**
39. Leaf terete, pendent ············· **35. Dendrobium (*D. vagans* only)**
    Leaf dorsiventrally flattened ··································· **40**
40. Stems superposed; leaves elliptic; lip apex fimbriate-lacerate ·· **37. Flickingeria**
    Stems clustered, not superposed; leaves linear; lip apex entire, ovate ·· **36. Diplocaulobium**
41. Flowers solitary or fasciculate in the axil of the terminal leaf, campanulate, red or yellow ································· **42**
    Flowers several to many, racemose or in an elongate inflorescence with fasciculate branches; flowers never red ······················ **43**
42. Flowers solitary or rarely in twos; rhizome creeping with well spaced pseudobulbs ································· **27. Mediocalcar**
    Flowers fasciculate; pseudobulbs superposed ·········· **28. Epiblastus**
43. Pollinia 6 ······································· **30. Appendicula**
    Pollinia 8 ··· **44**
44. Stem not pseudobulbous, elongate, strongly bilaterally compressed; inflorescence elongate, fasciculate; flowers lacking a mentum; lip bipartite, the base separated from the apex by a transverse callus ·· **29. Agrostophyllum**
    Stem pseudobulbous or not, short, not strongly bilaterally compressed; inflorescence racemose; laxly to densely many-flowered; flowers with a distinct mentum; lip entire or 3-lobed ···················· **44**

45. Flowers relatively large; sepals and petals more than 4 mm long **26. Eria**
    Flowers minute; sepals and petals 2.5 mm long or less · · · · · · · · · · **46**
46. Flowers with a column-foot and chin-like mentum
    · **31. Phreatia** (*P. micrantha* **only**)
    Flowers lacking a column foot and mentum · · · · · · · · · · · · · · · · **47**
47. Ovary winged · **32. Thelasis**
    Ovary not winged · **31. Phreatia**
48. Leaves terete or nearly so and grooved on upper surface · · · · · · · · **49**
    Leaves dorsiventrally flattened · **50**
49. Flowers racemose, tiny, less than 4 mm across; lip off-white, porrect,
    spurred at base · **41. Schoenorchis**
    Flowers in a sessile inflorescence, more than 8 mm long; lip deep
    purple, deflexed, lacking a spur · **42. Luisia**
50. Flowers fugacious, lasting a day or less · · · · · · · · · · **40. Thrixspermum**
    Flowers long-lasting, persisting for several days · · · · · · · · · · · · · · **51**
51. Plants tiny; largest leaves less than 3 cm long, deciduous, often absent at
    flowering; bracts with stipular outgrowths at base; lip shortly spurred,
    lacking any appendage within · · · · · · · · · · · · · · · · · · **43. Microtatorchis**
    Plants larger; largest leaves more than 5 cm long, persistent; bracts
    entire; lip saccate at base with a distinct appendage projecting from the
    posterior wall · **44. Pomatocalpa**

1. CORYMBORKIS

Thouars in Nouv. Bull. Sci. Soc. Phil. Paris 1: 318 (1809)

Corymborchis Blume, Fl. Javae Nov. Ser.: 105(1858–9).
Corymbis Thouars ex Rchb.f. in Seem., Fl. Vit.: 295(1868).

Large terrestrial plants with woody bamboo-like erect leafy stems growing from short rhizomes, often forming clumps; roots fibrous, fasciculate. Leaves in upper half of stem, plicate, ovate, elliptic or lanceolate. Inflorescences axillary, simple or branching. Flowers not opening widely, resupinate. Sepals and petals free or shortly connate, linear-oblanceolate, subsimilar. Lip parallel to column, spathulate, recurved at apex, with 2 longitudinal callus ridges. Column long, slender, clavate, straight; pollinia 2, granulose, with a slender terete stipe attached to a peltate viscidium; rostellum bilobed.

A pantropical genus of six or seven species, only one species found in Samoa.

C. veratrifolia (Reinw.)Blume, Fl. Javae, ser. 2,1, Orch.: 125 (1859). Type: Java, *Lobb* 162 (neo. K!).
For full synonymy see F. Rasmussen in Bot. Tidsskr. 71: 170 (1977).

Stems up to 1.2 m tall, leafy in apical half, often forming colonies. Leaves elliptic to lanceolate, acuminate, 20–32 × 4–9 cm, narrowed into a short petiole above the sheathing base. Inflorescences 1-several, up to 16 cm long, lateral from the leaf axils in upper part of stem, usually branching; bracts lanceolate, acuminate, 8–10 mm long. Flowers white or greenish white, narrow and tubular, not opening widely; pedicel and ovary up to 2 cm long. Sepals and petals similar, linear-oblanceolate, acute, 20–23 × 1.5–2 mm. Lip spathulate, 18–20 × 5–7 mm, claw linear, apical lamina subcircular-ovate, shortly apiculate, margins crispate. Column 15–16 mm long. Capsule stalked.

DISTRIBUTION: Savai'i and 'Upolu. Widely distributed from S.E. Asia and the Malay Archipelago to New Guinea, Australia, the Solomon Islands, Vanuatu, Fiji and Tonga.
HABITAT: Rare in montane rain-forest; 100–650 m.
COLLECTIONS: *Christophersen* 2604 (BISH, P), 3466 (BISH); *Graeffe* 1272 (BM, HAW, HBG); *Vaupel* 327(B†); *Whistler* 1034 (HAW, K), 1666 (BISH), 8232a (HAW); *Whitmee* 243 (K).

2. TROPIDIA

Lindl. in Bot. Reg. 19: sub t.1618 (1833).

Cnemidia Lindl. in Bot. Reg. 19: sub t.1618 (1833).

Medium-sized to rather large terrestrial plants with erect wiry and leafy stems, rarely saprophytic; roots fibrous, fasciculate. Leaves plicate or somewhat

so, lanceolate to linear-lanceolate, shortly petiolate, not articulated. Inflorescences terminal or axillary, simple or branching, few to many-flowered; bracts frequently distichous. Flowers usually small, up to 2 cm long, resupinate or non-resupinate. Sepals free, the laterals often connivent and embracing the base of the lip. Petals free, similar to the dorsal sepal. Lip usually larger than the other segments, sessile, more or less parallel to the column, entire or lobed, saccate, spurred or gibbous at the base, the disk with 2 longitudinal callus ridges. Column rather short; pollinia 2, granulose; stipe thin, terete; viscidium small, more or less peltate; rostellum bilobed.

A small genus of approximately 20 species in India, S.E. Asia, N.E. Australia and the S.W. Pacific Islands, and with two species in the tropical Americas. A single species is reported from Samoa.

T. effusa *Rchb.f.* in Seem., Fl. Vit.: 295 (1868). Type: Fiji, Taveuni, *Seemann* 612 (holo. K!, iso. W!)

Cnemidia ctenophora Rchb.f. in Otia Bot. Hamburg.: 51 (1878) & Xenia Orch. 3: 28 (1881). Type: Fiji, Ovalau, U.S. Expl. Exped., *Wilkes* s.n. (lecto. W!, isolecto. AMES!)

Tropidia ctenophora (Rchb.f.) Benth. & Hook.f. ex Drake, Ill. Fl. Ins. Mar. Pacif.: 311 (1892).

Plants with stems 20–75 cm tall. Leaves lanceolate, acuminate, 12–35 × 1.4–5.4 cm, very shortly petiolate above tubular sheathing base. Inflorescence simple or with a few branches, up to 12 cm long, many-flowered but flowers produced one at a time on each branch; bracts distichous, conduplicate, broadly ovate to lanceolate, acute, 7–12 mm long. Flowers glabrous, white or pale yellow, not opening widely; pedicel and ovary 7–8 mm long. Sepals ovate or lanceolate, acuminate, 7–10 × 2–2.5 mm. Petals oblong-lanceolate, acuminate, 6–8 × 1.5–2.5 mm. Lip oblong-ovate, acute and reflexed at apex, 5–7 × 2.5–3.5 mm, saccate at the base; callus of two ridges on the lateral veins in basal part and near apex. Column 2.5–3 mm long.

DISTRIBUTION: Savai'i. Also in Fiji.
HABITAT: Terrestrial orchid found rarely in montane forests; c.400 m.
COLLECTIONS: *Vaupel* 531 (B†); *Whistler* 39 (HAW); *Wilkes* s.n. (AMES, K, W).

NOTE: *Tropidia diffusa* Schltr. from New Guinea, the Solomon Islands and Vanuatu is similar and may prove to be conspecific, but it is a later name.

Fig. 1. *Corymborkis veratrifolia*. **A**, inflorescence and stem apex × $\frac{2}{3}$; **B**, habit × $\frac{1}{12}$; **C**, dorsal sepal × 1; **D**, lateral sepal × 1; **E**, petal × 1; **F**, lip × 1; **G**, column in side view × 1; **H, J**, column apex and anther cap × 4; **K**, pollinarium × 4. A drawn from plant cultivated by G. Dennis: **B–J** from *Wickison* s.n.. All drawn by Sue Wickison.

3. ERYTHRODES

Blume, Bijdr. Fl. Ned. Ind.: 410 (1825).

Physurus Rich., Orch. Eur. Annot. 33 (1817) nom. nud.; Rich. ex Lindl., Gen. Sp. Orch. Pl.: 501(1840).

Small to medium-sized terrestrial plants with fleshy creeping rhizomes, rooting at the nodes, and erect leafy fertile shoots. Leaves not articulate, thin-textured. Inflorescence terminal, erect, racemose, many-flowered. Flowers small, not opening widely, resupinate. Dorsal sepal forming with the petals a hood over the column. Lateral sepals spreading. Petals entire, membranous, narrower than the sepals. Lip appressed to the column, entire or with a small apical recurved blade, prolonged at the base into a spur that projects between the lateral sepals; spur entire or bilobed, short to long, with 2–4 sessile calli within. Column short; anther erect; pollinia 2, sectile, clavate, attached by caudicles to a small viscidium; rostellum short, bilobed.

A genus of about 100 species widespread in the tropics of the New World, tropical Asia and the Pacific region but absent from Africa. Two species have been recorded in Samoa.

Sepals 6–8 mm long, densely hairy; lip apex acute; leaves 5–10 cm long
· **E. oxyglossa**
Sepals 2.5–4 mm long, glabrous or sparsely hairy on outer surface; lip apex a
small ovate, obtuse blade; leaves 3–5.5 cm long · · · · · · **E. purpurascens**

E. oxyglossa *Schltr.* in Bot. Jahrb. Syst. 39: 53 (1906). Type: New Caledonia, *Schlechter* 15749 (holo. B†)

Physurus lilyanus H.Fleischm. & Rech. in Denkschr. Kaiserl. Akad. Wiss., Math.-Naturwiss. Kl. 85: 253, t.1, fig.3 (1910). Types: Samoa, *Rechinger* 63 (syn. W!); *Rechinger* 1590 (syn. W!); *Rechinger* 1515 (syn. W!).

Erythrodes lilyana (H.Fleishm. & Rech.) Schltr. in Repert. Spec. Nov. Regni Veg. 9: 87 (1910).

Plant 25–40 cm tall. Leaves obliquely oblong-lanceolate to oblong-ovate, acuminate, 5–10 × 1.5–2.5 cm; petiole slender, 2.5–3.3 cm long. Inflorescence 12–27 cm long, many-flowered; peduncle pubescent in upper part; rachis pubescent; bracts lanceolate, 8–10 mm long, hairy. Flowers dull brownish white, pubescent on the outside of the sepals. Sepals lanceolate, 6–8 × 7–2.3 mm. Petals obliquely oblanceolate, acute, 5.5–7.5 × 1.5–1.7 mm. Lip 6–8.5 mm long, lanceolate to oblong-lanceolate, acute; spur bilobed at apex, 2.5–3.5 mm long, with 2 small glands within. Column 2.5–3.5 mm long.

DISTRIBUTION: Olosega, Savai'i, Ta'u, 'Upolu. Also in New Caledonia, Fiji and Tonga.

HABITAT: Terrestrial orchid found in montane and cloud forest; 300–1300 m.

COLLECTIONS: *Rechinger* 63 (W), 1515 (W), 1590 (W); *Vaupel* 654 (B†);

Whistler 40 (HAW), 560 (HAW), 593a (HAW), 1655 (HAW), 1764 (HAW), 1908b (HAW), 1980 (HAW), 3100 (HAW), 4775 (HAW), 5006 (HAW), 6893 (HAW), 7799 (HAW), 8012 (HAW), 8354 (HAW), 9908 (HAW).

E. purpurascens *Schltr.* in K.Schum. & Lauterb., Nachtr. Fl. Schutzgeb. Sudsee 88 (1905). Type: Papua New Guinea, nr. Paub, *Schlechter* 14616 (holo. B†). *Cheirostylis sp.* sensu Yuncker in Bernice P. Bishop Mus. Bull. 184: 31 (1945).

Plant up to 32 cm tall. Leaves obliquely ovate, acute, 3–5.5 × 1.5–2 cm. Inflorescence up to 22 cm long; bracts lanceolate, acuminate, 5–8 mm long. Flowers white, small, 4–6 mm long; pedicel and ovary 6–8 mm long, pubescent. Sepals elliptic-lanceolate, 2.5–4 × 1 mm, glabrous or very sparsely pubescent. Petals obliquely oblanceolate, acute, 2.5–3.5 × 0.7–0.8 mm. Lip subpandurate, 2.5–4.5 mm long, the apical lamina ovate, acute; spur 1.5–2.5 mm long, bilobed with 2 small calli within.

DISTRIBUTION: Ta'u, Tutuila, 'Upolu. Also in New Guinea and possibly Fiji and Tonga.
HABITAT: Uncommon in montane forest; 300–890 m.
COLLECTIONS: *Whistler* 2741 (BISH), 3598 (HAW), 3713 (HAW), 4388 (HAW); *Yuncker* 9265 (BISH).

NOTE: We have assigned these collections to *E. purpurascens* rather than to *E. parvula* Kores because of the presence of calli in the spur of the lip. *E. parvula* lacks these. The possibility that the latter may be a semi-peloric form of *E. purpurascens* needs further examination.

4. GOODYERA*
R.Br. in W.Aiton & W.T.Aiton, Hort. Kew ed.2, 5: 197 (1813).

Terrestrial herbs with creeping fleshy rhizomes, rooting at the nodes, and erect fleshy leafy fertile shoots. Leaves thin-textured to fleshy, not articulated. Inflorescence terminal, racemose, few to many-flowered. Flowers small, resupinate, usually tubular but never opening widely. Dorsal sepal and petals connivent into a hood over the column. Lateral sepals porrect, reflexed or spreading somewhat. Lip parallel to column, entire, deeply saccate at the base, with numerous thread-like calli within. Column short to long; anther erect, persistent; pollinia 2, sectile, clavate, often deeply divided, attached by caudicles to a rather large viscidium; rostellum long, bilobed.

Rechinger 1588, from Mauga-afi at 1200 m on Savai'i and misidentified by Fleischmann & Rechinger (1910) as *Eucosia carnea* Blume, has peloric flowers with a petaloid, ecallose, lanceolate lip. In column structure it appears to be referable to *Goodyera*, close to *G. viridiflora* Blume, but it has more oblong-ovate leaves with short petioles and obtuse sterile bracts on the peduncle. From the single specimen available it is impossible to place it in any of the known Pacific Island species of the genus.

A genus of about 40 species found in both temperate and tropical parts of the Old and New Worlds but not in Africa. A single species has been reported from Samoa.

G. rubicunda *(Blume) Lindl.* in Bot. Reg. 25: 61, misc. 92 (1839). Type: Java, *Blume* s.n. (holo. L!, iso. P!).
Neottia rubicunda Blume, Bijdr. Fl. Ned. Ind.: 408 (1825).
N. grandis Blume, l.c.: 407 (1825).
Goodyera grandis (Blume) Blume, Fl. Javae ser.2,1, Orch.:36 (1858).
G. rubens Blume, Fl. Javae ser.2, 1, Orch.: 36, t.9 (1858).
G. biflora sensu Kraenzl. in Bot. Jahrb. Syst. 25: 600 (1898), non Hook.f.
G. sp. sensu Kraenzl. *loc. cit.*
G. triandra Schltr. in Bull. Herb. Boissier 2, 6: 298 (1906). Type: New Hebrides (Vanuatu), *Morrison* s.n. (holo. B †; iso. AMES!).
G. anomala Schltr. in Repert. Spec. Nov. Regni Veg. 9: 86 (1910). Type: Samoa, *Vaupel* 405 (syn. B†), *Rechinger* 94) (syn. W!)
G. waitziana sensu H.Fleischm. & Rech. in Denkschr. Kaiserl. Akad. Wiss., Math.-Naturwiss. Kl. 85: 254 (1910), non Blume.
Goodyera rubicunda (Blume) Lindl. var. *triandra* (Schltr.)N.Hallé in Fl. Nouv.-Caled. Depend. 8: 532 (1977).

Plants up to 70 cm tall. Leaves obliquely elliptic to elliptic-lanceolate, acute or acuminate, 12–18 × 3.5–6 cm, long-petiolate. Inflorescence up to 40 cm long, pubescent. Flowers orangish, reddish brown or greenish pink. Sepals narrowly ovate to ovate, acute, 7–8.5 × 2.5–4 mm, densely pubescent. Petals clawed, obliquely ovate to subrhombic, acute, 7–7.5 × 2.5–3 mm. Lip entire, 6–7.5 mm long, saccate at base, recurved and ligulate at apex. Column 5–6.5 mm long.

DISTRIBUTION: Savai'i, 'Upolu, Ofu. Also from southern Ryukyu Islands and Malaya, throughout Malesia to northern Australia and eastwards to the Solomon Islands, Vanuatu, Fiji and Tonga.
HABITAT: Terrestrial orchid in lowland and montane forests; 200–900 m.
COLLECTIONS: *Bryan* 162 (BISH, K); *Christophersen* 2700 (BISH); *Graeffe* s.n. (W); *Powell* 269 (K); *Rechinger* 94 (W), 652 (W), 1556 (W), 1779 (W), 2506 (BM, W); *Reinecke* 273 (B†), 273a (B†), 291 (B†); *Vaupel* 405 (B†); *Whistler* 1765 (HAW), 3021 (BISH, K), 5005 (HAW); Wilkes s.n. (W).

5. **MOERENHOUTIA**
Blume, Fl. Javae ser.2,1, Orch.: 99 (1859).
Coralliokyphos H. Fleischm. & Rech. in Denkschr. Kaiserl. Akad. Wiss., Math.-Naturwissen. Kl. 85: 252 (1910)

Medium-sized terrestrial herbs with creeping fleshy rhizomes, rooting at the nodes, and erect fleshy fertile leafy shoots. Leaves thin-textured, several.

Inflorescence terminal, racemose, laxly to densely many-flowered, pubescent. Flowers not opening widely, ovary shortly but densely pubescent. Sepals fleshy; dorsal sepal adnate to the lateral sepals; lateral somewhat oblique, spreading. Petals thin-textured. Lip entire, parallel to the column, saccate at base, ovate, suborbicular or bilobed at apex; calli basal in saccate part of lip, comprising two bunches of short appendages.

A small genus of about a dozen species in New Guinea and the S.W. Pacific Islands. A single species reported from Samoa.

M. heteromorpha (*Rchb.f.*) *Benth. & Hook.f. ex Drake* in Ill. Ins. Mar. Pacif., Fasc. 7: 313 (1892). Types: Samoa, "Tuticella" [Tutuila] & 'Upolu, *Wilkes* s.n. (syn. W!).
Platylepis heteromorpha Rchb.f. in Otia Bot. Hamburg.: 52 (1878).
Coralliokyphos candidissimum H.Fleishm. & Rech. in Denkschr. Kaiserl. Akad. Wiss., Math.-Naturwissen. Kl. 85: 252 (1910). Type: Samoa, *Rechinger* 707 (holo. W!, iso. BM!).

Plant 25–50 cm tall. Leaves lanceolate or obliquely elliptic-ovate, acuminate, up to 16 × 6 cm. Inflorescence 13–25 cm long; peduncle and rachis shortly pubescent. Flowers white, pubescent on ovary and base and apex of sepals. Sepals fleshy, lanceolate, 9–11 mm long, pubescent at base and apex. Petals oblanceolate, one-veined, 9–9.5 × 2.5–3 mm. Lip oblong-ovate, 8–9 × 4–5 mm, with 3 central longitudinal fleshy ridges; callus of two basal tufts of appendages. Column 7–7.5 mm long.

DISTRIBUTION: Olosega, Savai'i, Ta'u , Tutuila, 'Upolu. Possibly endemic.
HABITAT: Terrestrial orchid occasionally found in lowland and montane forest; 300–750 m.
COLLECTIONS: *Christophersen* 1822 (BISH); *Cox* 198 (BISH); *Garber* 1073 (BISH); *Graeffe* 1270 (HBG); *Rechinger* 707 (W); *Vaupel* 406 (B†); *Whistler* 421 (BISH, HAW), 1979 (HAW), 2685 (BISH, HAW), 2729 (HAW), 2775 (BISH, HAW), 3097 (BISH, HAW), 3172 (BISH, HAW), 3583 (HAW), 7071 (HAW), 7088 (HAW), 8014 (HAW), 8313 (BM, HAW).

NOTE: Very closely allied to *M. plantaginea* Blume from the Society Islands and *M. grandiflora* Schltr. from New Caledonia and Vanuatu. The former appears to have slightly smaller flowers and a shorter stem, but their relationship needs further investigation. The latter differs slightly in having petals with a branching venation and a lip lacking a prominent 3-ridged callus.

6. **PRISTIGLOTTIS**

Cretz. & J.J.Sm. in Acta Fauna Fl. Universali ser.2, Bot.1 (14): 4 (1934).

Small terrestrial plants with creeping horizontal fleshy rhizomes, rooting at the nodes. Stems erect, leafy in lower part. Leaves alternate, spreading, petiolate, not articulated. Inflorescence terminal, racemose, few-flowered.

Flowers small to medium-sized, white or off-white, tubular, resupinate. Dorsal sepal adnate to the petals to form a hood over the column. Lateral sepals similar to the dorsal but obliquely extended at the base around the saccate base of the lip. Petals simple, membranous. Lip tripartite, the base saccate enclosing two small sessile glands, the middle an elongate channelled claw with undulate-crenate margins, the apical part a flat lamina. Column elongate, with two parallel elongate appendages on the ventral surface projecting into the saccate base of the lip: anther erect, persistent; pollinia two, sectile, clavate; viscidium solitary; rostellum narrowly attenuate or linear-ligulate, deeply bilobed.

A small genus of perhaps 15 species from India and S.E. Asia to the Malay Archipelago and S.W. Pacific Islands. A single species recorded from Samoa.

P. longiflora *(Rchb.f.) Kores* in Allertonia 5(1): 28 (1989). Type: Fiji, Taveuni, *Seemann* 601 (holo. W!, iso. BM!, K!).

Anecochilus longiflorus Rchb.f. in Seem., Fl. Vit.: 294 (1868).

Odontochilus longiflorus (Rchb.f.)Benth. & Hook.f. ex Drake in Ill. Fl. Ins. Mar. Pac.: 312 (1892).

Cystopus funkii Schltr. in Repert. Spec. Nov. Regni Veg. 9: 89 (1910). Types: Samoa, 'Upolu, *Funk* s.n. (syn. NSW), *Vaupel* 543, *Reinecke* 217, *Hochreutiner* 3267 (all syn. B†, isosyn. L!, Z!).

Odontochilus upoluensis Kraenzl. in Mitt. Inst. Allg. Bot. Hamburg 5: 236 (1922). Type: Samoa, *Graeffe* 1258 (holo. HBG!, iso. HAW!).

Pristiglottis funkii (Schltr.)Cretz. & J.J. Sm. in Acta Fauna Fl. Universali 2, Bot. 1(14): 4 (1934).

Cheirostylis longiflora (Rchb.f.) L.O.Williams in Bot. Mus. Leafl. 7: 138 (1939).

A small terrestrial herb 8–20 cm tall. Leaves ovate, acute, 2.5–6.5 cm long, 1–2.5 cm wide, petiolate and sheathing at base. Inflorescence 5–10 cm long, 2–10-flowered. Flowers white; pedicel and ovary c. 1 cm long. Dorsal sepal linear-lanceolate, subacute, 13–20 × 3–4 mm, glabrous or sparsely hairy. Lateral sepals similar but with an obliquely extended base. Petals ligulate, obtuse, 11–18 × 1.5–2.5 mm, adnate to the dorsal sepal. Lip 15–20 mm long, saccate at base, the claw relatively slender, channelled, 8–10 mm long, crenulate on lateral margins, the apical lamina entire, elliptic or obovate-elliptic, obtuse, 5–8 × 3–5 mm. Column slender 8–10 mm long.

DISTRIBUTION: Savai'i, 'Upolu. Also found in the Solomon Islands, Vanuatu and Fiji.

HABITAT: Ground orchid occasionally found in montane scrub and montane forest; 660–900 m.

COLLECTIONS: *Bryan* 101 (BISH), 172 (BISH); *Christophersen* 2161 (BISH); *Funk* s.n. (NSW); *Graeffe* 1258 (HAW, HBG); *Hochreutiner* 3267 (G); *Reinecke* 217 (B†), 291(B†); *Vaupel* 543 (B†); *Whistler* 424 (HAW), 561 (HAW), 1756 (HAW); *Wilkes* s.n.(W).

7. ANOECTOCHILUS

Blume, Fl.Javae, Praef.: 6, in adnot. (1828).

Terrestrial plants with creeping fleshy rhizomes, rooting at the nodes. Erect stems leafy. Leaves entire, not articulated, fleshy-membranous, often dark blackish maroon marked on the veins with red, gold or silver, purple beneath. Inflorescence terminal, laxly few-flowered, racemose. Flowers quite large for plant, showy. Dorsal sepal concave, adnate to the membranous petals to form a hood over the column. Petals entire. Lip tripartite; base saccate or spurred, bearing stalked glands within; middle part claw-like, channelled, with lacerate margins; apical part a transverse lamina. Column short with two parallel lamellate appendages on ventral surface; anther erect, persistent; pollinia two, clavate, sectile, joined to a common viscidium.

A genus of some 50 species distributed from India throughout tropical East Asia to the Malay Archipelago, the Philippines and the S.W.Pacific Islands. A single species in Samoa.

A. imitans *Schltr.* in Bot. Jahrb. Syst. 39: 54 (1906). Type: New Caledonia, near Paita, *Schlechter* 14864 (holo. B†).
A. vitiensis Rolfe in J. Linn. Soc. Bot. 39: 176 (1909). Type: Fiji, Viti Levu, *Gibbs* 635 (holo. BM!, iso. K!).

Plant 5–20 cm tall. Leaves 3–6, broadly ovate, mucronate, 1.5–4.5 × 2–3.5 cm, dark iridescent blackish green, with reddish or gold venation; petiole 1–2 cm long. Inflorescence 2–7-flowered, up to 15 cm long; peduncle and rachis pubescent. Flowers white or greenish white, villose on outer surface of the sepals, pedicel and ovary. Dorsal sepal ovate-lanceolate, acuminate, 11–13 × 4–5 mm. Lateral sepals obliquely oblong-lanceolate, acuminate, 11–12 × 3.5–4 mm. Petals narrowly oblong-lanceolate, acuminate, 10–11 × 2.5–3 mm. Lip 14–17 mm long; basal and middle parts 6–7 mm long; apical lamina oblong-lanceolate, 5–6 × 2–3 mm; spur conical, obtuse, 4–5 mm long, bearing two oblong, papillose glands. Column 8–9 mm long.

DISTRIBUTION: Samoa, without exact locality, New Caledonia and Fiji.
HABITAT: In Fiji in dense forest from 150–900 m.
COLLECTION Herb.Mueller (MEL).

NOTE: The only Samoan collection of this species is sterile but it matches closely material from Fiji and New Caledonia we have seen. It is distinctly possible that the attribution of this specimen to Samoa is erroneous.

8. HETAERIA

Blume, Bijdr. Fl. Ned. Ind.: 409 (1825).

Small to medium-sized terrestrial herbs with fleshy creeping rhizomes, rooting at the nodes. Stems erect, leafy. Leaves simple, membranous, petiolate and sheathing at the base. Inflorescence terminal, racemose, few to many-flowered, lax or dense. Flowers small, non-resupinate. Dorsal sepal entire, adnate to the membranous petals. Lateral sepals similar to the dorsal but oblique, enclosing the saccate lip-base. Lip more or less parallel to the column, entire, the base saccate, bearing two or more unstalked papillae or glands within, the apex small, concave, often contracted into a claw with a terminal bilobed transverse blade. Column short, with two parallel lamellae or keels on ventral surface; anther persistent, erect; pollinia two, sectile, attached to a common viscidium; rostellum short, bilobed.

A genus of perhaps 70 species widespread from India throughout S.E. Asia and the Malay Archipelago to northern Australia and the S.W. Pacific Islands. Also in Africa and Madagascar and possibly the New World tropics. Two species reported from Samoa.

Leaves elliptic to ovate-elliptic, 4–12 cm long, 2–4.5 cm wide; lip apex very
    small, not expanded into a transverse blade · · · · · · · · · · **H. oblongifolia**
Leaves lanceolate, 10–22 cm long, 0.7–2.5 cm wide; lip apex expanded into a
    transversely bilobed blade · **H. whitmeei**

H. oblongifolia *Blume*, Bijdr. Fl. Ned. Ind.: 410 (1825). Type: Java, *Blume* s.n. (holo. L!).
Aetheria oblongifolia (Blume) Lindl., Gen. Spec. Orch. Pl.: 491 (1840).
Rhamphidia discoidea Rchb.f. in Linnaea 41: 59 (1877).Type: New Caledonia, *Vieillard* 1311 (holo. P!, iso. BM!).
Hetaeria forcipata Rchb.f. in Linnaea 41: 62 (1877). Type: Fiji, *Roezl* s.n. (holo.W).
Hetaeria samoensis Rolfe in Bull. Misc. Inform., Kew 1898: 199 (1898). Type: Samoa, *Walter* s.n. (holo. CAM!, iso. K!).
Goodyera discoidea (Rchb.f.) Schltr. in Bot. Jahrb. Syst. 39: 57 (1906).
Hetaeria similis Schltr. in Repert. Spec. Nov. Regni Veg. 9: 88 (1910). Type: Samoa, *Vaupel* 657 (holo. B†, iso. BISH!, K!).
H. discoidea (Rchb.f.)Schltr., l.c.: 89 (1910).
Rhamphidia tenuis sensu H.Fleischm. & Rech. in Denkschr. Kaiserl. Akad. Wiss., Math.-Naturwiss. Kl. 85: 254 (1910), non Lindl.
Hetaeria raymundi Schltr. in Bot. Jahrb. Syst. 56: 453 (1921). Type: Palau Islands, *Raymundus* s.n. (syn. B†), *Ledermann* 14312 (syn. B†).

Plant up to 50 cm tall. Leaves obliquely elliptic to ovate-elliptic, acute or abruptly acuminate, 4–12 × 2–4.5 cm; petiole 1.5–4 cm long. Inflorescence 10–30 cm long, many-flowered. Flowers small, white or creamy white; pedicel and ovary 5–7 mm long. Dorsal sepal ovate, subacute, 3–4 × 1.25–1.75 mm. Lateral sepals obliquely oblong-ovate, obtuse, 3–4 × 1–1.25 mm. Petals narrowly elliptic, obtuse, 3–3.5 × 0.7–1 mm. Lip saccate, 3–3.5 × 1 mm, mucronate at apex; calli papillate, 2 or 3 pairs. Column 1–1.5 mm long.

DISTRIBUTION: Ofu, Olosega, Savai'i, Ta'u, 'Upolu. Also widely distributed from the Philippines and Indonesia, through New Guinea and the Palau Islands, to the Solomon Islands, Santa Cruz Islands, Vanuatu, New Caledonia, Fiji and Australia.

HABITAT: Terrestrial orchid found in rain-forests; 250–600 m.

COLLECTIONS: *Christophersen* 2627 (BISH, K), 2699?; *Rechinger* 102 (W), 1147 (W), 1662 (W), 1884 in part (BM, W); *Reinecke* 34 (K); *Vaupel* 657 (B, BISH, K); *Walter* s.n. (CAM, K); *Whistler* 826 (HAW), 3022 (HAW, K), 3084 (BISH, HAW, K), 3165 (BISH, HAW, K), 3916 (BISH, HAW), 6930 (HAW).

H. whitmeei *Rchb.f.* in J. Bot. 15: 133 (1877). Type: Samoa, *Whitmee* s.n. (holo. K!, iso. W!).

Hetaeria polyphylla Rchb.f. in Otia Bot. Hamburg. (as *Etaeria*): 52 (1878) & Xenia Orch. 3: 29 (1881). Type: Vanua Levu, *U.S. Expl. Exped.* (holo. W!, iso. and photograph at AMES).

Adenostylis stricta Rolfe in J. Linn. Soc. Bot. 39: 177 (1909). Type: Viti Levu, *Gibbs* 667 (holo. BM!).

Zeuxine sphaerocheila H.Fleischm. & Rech. in Denkschr. Kaiserl. Akad. Wiss. , Math.-Naturwiss. Kl. 85: 251, t.2, fig.6 (1910). Types: Samoa, *Rechinger* 3710 (syn. W!) & *Rechinger* 1663 (syn. W!).

Zeuxine betchei Schltr. in Repert. Spec. Nov. Regni Veg. 9: 90 (1910). Type: Samoa, *Betche* 57 (holo. B†, iso. MEL!).

Hetaeria francisii Schltr. in Repert. Spec. Nov. Regni Veg. 9: 161 (1910). Type: New Caledonia, *Francis* 767 (holo. B†, iso. P!, ex herb. Bonati).

Zeuxine francisii (Schltr.)Schltr. in Repert. Spec. Nov. Regni Veg. 9: 289 (1911).

Zeuxine triandra Hotta in Acta Phytotax. Geobot. 19: 156 (1963). Type: Tonga, *Hotta* 5332 (holo. KYO).

Plants up to 60 cm tall. Leaves narrowly elliptic to lanceolate, acuminate, 10–22 × 0.7–2.2 cm; petiole 2–3.5 cm long. Inflorescence glabrous, 20–40 cm long, laxly many-flowered. Flowers brownish white without, creamy within. Dorsal sepal ovate, subacute, 4–5 mm long. Lateral sepals obliquely oblong-ovate, subacute, 4–5.5 × 2–2·5 mm. Petals oblong, rounded at apex, 4–4.5 mm long, 0.7–1 mm wide. Lip 4.5–5.5 × 3–4 mm, saccate at the base with two lamellate papillate calli within and longitudinally divided by a shallow groove externally; apical part transversely oblong, 1–1.5 × 1.5–2 mm. Column 1.5 mm long.

DISTRIBUTION: Savai'i, Ta'u, 'Upolu. Also in New Caledonia, Fiji and Tonga.

HABITAT: Terrestrial in lowland and montane forest; 200–650 m.

COLLECTIONS: *Betche* 57 (MEL); *Christophersen* 542 (BISH), 653 (BISH); *Graeffe* 1268 (HBG); *Rechinger* 1663 (W), 3710 (W); *Whistler* 3696 (HAW); *Whitmee* s.n. (K, W).

9. **VRYDAGZYNEA**

Blume, Fl. Javae ser. 2,1, Orch.: 59 (1858).

Small terrestrial herbs with creeping fleshy rhizomes, rooting at the nodes. Erect stems leafy. Leaves entire, alternate, petiolate, sheathing at the base, membranous. Inflorescence terminal, racemose, laxly to densely many-flowered. Flowers small, resupinate. Dorsal sepal entire, adnate to the petals and forming a hood over the column. Lateral sepals similar to the dorsal but oblique at the base. Petals entire, membranous. Lip more or less parallel to the column, entire, prominently spurred at the base, the spur projecting between the lateral sepals, entire or bilobed at apex, with two stalked glands within. Column very short; anther erect, persistent; pollinia two, sectile, clavate, attached to a common viscidium; rostellum relatively short, bilobed.

A small genus of perhaps 20 species in India, S.E. Asia, the Malay Archipelago, northern Australia and the S.W. Pacific Islands. Two species have been recorded in Samoa.

Leaves green; lateral sepals hooded at apex; spur longer than the sepals
· **V. samoana**
Leaves dark green above with a central longitudinal white stripe; lateral
sepals not hooded at apex; spur shorter than the sepals · · · · **V. vitiensis**

V. samoana *Schltr.* in Repert. Spec. Nov. Regni Veg. 9: 91 (1910). Types: Savai'i, *Rechinger* 1145 (syn. B†, isosyn. BM!, US, W!) & *Vaupel* 655 (syn. B†).
Vrydagzynea albida (Blume) Blume var. *purpurascens* Kraenzl. in Bot. Jahrb. Syst. 25: 599 (1898). Type: Savai'i, *Rechinger* 542 (holo. B†, iso. W!).
V. whitmeei sensu H.Fleischm. & Rech. in Denkschr. Kaiserl. Akad. Wiss., Math.-Naturwiss. Kl. 85: 253 (1910) (as *Vrydagzynea whitmeei*), non Schltr.

Plant 10–20 cm tall. Leaves narrowly ovate to elliptic-ovate, acuminate, 2.5–5.5 × 1–2 cm, light to dark green; petiole 1–1.5 cm long. Inflorescence 6–10 cm long; peduncle, glabrous or sparsely pubescent near apex, longer than the rachis; bracts lanceolate, 5–8 mm long. Flowers small, pale green to greenish white, occasionally with a pink tinge; pedicel and ovary c. 1 cm long. Dorsal sepal ovate, obtuse, 3–3.5 × 1.7–2.2 mm. Lateral sepals obliquely oblong-ovate, obtuse, 2.7–3.5 × 1.5–2.2 mm, carinate on outside. Petals obliquely oblong-ovate, subacute, 2.5–3 × 1.7–2 mm. Lip oblong-suborbicular, obtuse, 2–2.5 mm long, 2 mm wide; spur cylindrical-subfusiform, 4–5 mm long, obscurely bilobed at apex. Column 1–1.5 mm long.

DISTRIBUTION: Savai'i, Ta'u, 'Upolu. Also in Vanuatu and Fiji.
HABITAT: Terrestrial orchid found in montane and cloud-forests; 175–650 m.
COLLECTIONS: *Christophersen* 871 (BISH, K); *Garber* 630 (BISH); *Rechinger* 1145 (BM, W), 1699 (W), 1804 (W), 5318 (W); *Reinecke* 542 (B†); *Vaupel* 655 (B†); *Whistler* 787 (HAW), 1512 (HAW), 1908a (HAW, K), 4774 (HAW, K).

V. vitiensis Rchb.f., Otia Bot. Hamburg.: 51 (1878). Types: Samoa, *Wilkes* s.n. (syn.W!) & Fiji, Ovalau, *Wilkes* s.n. (lecto. W!, isolecto. AMES!).
V. whitmeei Schltr. in Bull. Herb. Boissier ser. 2, 6: 296 (1906). Type: Samoa, *Whitmee* s.n. (holo. B†).
V. sp. sensu Yuncker in Bernice P. Bishop Mus. Bull. 184: 31 (1945).

Plants 8–25 cm tall. Leaves lanceolate or narrowly ovate, acute, 3.5–6 × 1.2–2 cm, green with a central white stripe; petiole 0.8–1.5 cm long. Inflorescence densely many-flowered, 2–7 cm long; rachis short, sparsely villose. Flowers greenish white to white; pedicel and ovary 5–9 mm long. Dorsal sepal oblong-ovate, subacute, 3–3.5 × 1–1.5 mm. Lateral sepals oblong, obtuse, oblique at the base, 3–3.5 × 1.5–2 mm. Petals falcately linear-oblong, obtuse, 2–2.5 × 1–1.5 mm. Lip ovate, blunt with incurved sides, 2–2.5 × 1–1.7 mm; spur conical, obscurely bilobed at apex, 2.5–3 mm long. Column 1.5–2 mm long.

DISTRIBUTION: Savai'i, Ta'u, 'Upolu. Also in Vanuatu and Fiji.
HABITAT: Terrestrial orchid found in rain- and cloud-forests; 250–900 m.
COLLECTIONS: *Christophersen* 253 (BISH, P), 3187 (BISH), 3250 (BISH); *Lister* s.n. (K); *Rechinger* 1145 (W), 1804 (W), 5318 (W); *Vaupel* 517 (B†); *Whistler* 290 (HAW), 511 (HAW), 3553 (HAW, K), 3913 (BISH, HAW), 3982 (HAW), 4005 (BISH, HAW), 4776 (HAW), 6890 (HAW); *Yuncker* 9269 (BISH); *Whitmee* s.n. (K).

10. ZEUXINE

Lindl., Coll. Bot., app. no.18 (1826).

Adenostylis Blume, Bijdr. Fl. Ned. Ind.: 414 (1825)
Monochilus Wall. ex Lindl., Gen. Spec. Orch. Pl.: 486 (1840).

Small to medium-sized terrestrial herbs with creeping fleshy rhizomes rooting at the nodes. Erect stems leafy, fleshy. Inflorescences terminal, racemose, few- to many-flowered. Flowers small, resupinate. Dorsal sepal entire, adnate to the petals to form a hood over the column. Lateral sepals free, similar to dorsal sepal but oblique at base. Lip more or less adnate to front of column, entire, saccate or cymbiform at the base, with two small glands within; the middle clawed, the apex broadened into a transverse blade. Column short; anther dorsal, erect, persistent; pollinia two, sectile, attached to a common rather large viscidium; rostellum prominent, rather broad, deeply divided.

A large genus of perhaps 70 species widespread in the Old World tropics from Africa, Madagascar and India across to China, S.E. Asia, the Malay Archipelago and the Pacific Islands.

1. Flowers glandular; sepals and petals 5.5 mm long or more; lip with a central longitudinal keel in the basal saccate part; leaves more than 5.5 × 2.4 cm ···**Z. plantaginea**

Flowers sparsely pubescent; sepals and petals 4 mm long or less; lip
lacking a longitudinal keel in the basal part · · · · · · · · · · · · · · · · · · **2**
2. Lip with basal part longer than the apical lamina; leaves lanceolate, 1.3
cm wide or less · **Z. stenophylla**
Lip with basal part shorter than the apical lamina; leaves ovate, more
than 1.6 cm wide · **Z. vieillardii**

Z. plantaginea *(Rchb.f.) Benth. & Hook.f. ex Drake*, Ill. Fl. Ins. Mar. Pacif.: 312
(1892). Type: Samoa, U.S. Expl. Exped., *Wilkes* s.n. (holo. W!).
Monochilus plantagineus Rchb.f., Otia Bot. Hamburg.: 52 (1878).
Zeuxine androcardium Schltr. in Repert. Spec. Nov. Regni Veg. 9: 90 (1910).
Type: Samoa, *Vaupel* 287 (holo. B†, iso. AMES!, K!).

Plants 25–35 cm tall. Leaves elliptic-ovate or obliquely elliptic-ovate, acute,
5.5–8.5 × 2.5–3.5 cm; petiole 2–3 cm long. Inflorescence 10–15-flowered;
peduncle and rachis hairy; bracts lanceolate, acuminate, 6–12 mm long.
Flowers small, white; pedicel and ovary 5–8 mm long, glandular-pubescent.
Dorsal sepal deeply cucullate, ovate, obtuse, 5–5.5 × 3–3.5 mm. Lateral sepals
obliquely ovate, obtuse, 5–6 × 3 mm. Petals obliquely ovate, subacute, 5–5.5 ×
3–3.5 mm. Lip 5 × 3 mm, saccate at base, transversely oblong at apex; callus of
two digitate flanges in saccate part of lip. Column short, 3–3.5 mm long.

DISTRIBUTION: Olosega, Savai'i, 'Upolu. Endemic.
HABITAT: Terrestrial orchid found in rain-forests and cloud-forests;
300–800 m.
COLLECTIONS: *Christophersen* 38 (BISH, P), 875 (BISH), 2295 (BISH, K),
2740 (BISH); *Vaupel* 287 (AMES, K); *Whistler* 3100 (HAW), 3974 (BISH, HAW),
4006 (BISH, HAW, K), 5931 (HAW, K); *Wilkes* s.n. (W).

Z. stenophylla *(Rchb.f.) Benth. & Hook.f. ex Drake*, Ill. Fl. Ins. Mar. Pacif.: 312
(1892).Type: Samoa, Savai'i & Tutuila, *Wilkes* s.n. (syn. W!).
Monochilus stenophyllus Rchb.f. in Otia Bot. Hamburg.: 52 (1878) & Xenia
Orch. 3: 29 (1881).
Adenostylis vitiensis Rolfe in J. Linn. Soc. Bot. 39: 177 (1909). Type: Viti Levu,
Gibbs 618 (holo. BM!, iso. K!).
Zeuxine vitiensis (Rolfe) L.O. Williams in Bot. Mus. Leafl. 5: 112 (1938).

Plants 18–35 cm tall. Leaves lanceolate, acute, 1.8–5 × 0.6–1.3 cm.
Inflorescence laxly many-flowered; peduncle and rachis sparsely hairy; bracts
lanceolate, acuminate up to 8 mm long. Flowers white, sparsely hairy; pedicel
and ovary 5–7 mm long, sparsely hairy. Dorsal sepal ovate, obtuse, 3–4 × 2–3
mm. Lateral sepals obliquely oblong-ovate, obtuse, 4–5 × 2–3 mm. Petals
obliquely ovate, obtuse, 3–3.5 × 2–2.5 mm. Lip 3.5–4 × 2 mm, recurved at
apex; basal part bearing two recurved hook-like calli within; apical lamina
transversely oblong, 1 mm long, 2–2.5 mm wide. Column 1.7–2 mm long.

DISTRIBUTION: Savai'i, Tutuila, 'Upolu. Also found in Vanuatu, Fiji and Tonga.

HABITAT: In dense forest at elevations of 150–700 m.

COLLECTIONS: *Christophersen* 81 (BISH, K); *Rechinger* 1825 (W); *Whistler* 559 (HAW), 593 (HAW), 3974 (HAW), 4023 (HAW), 6934 (BISH, HAW) *Wilkes* s.n. (W).

Z. vieillardii *(Rchb.f.) Schltr.* in Bot. Jahrb. Syst. 39: 55 (1906).Type: New Caledonia, *Vieillard* 1311 (holo. P!, iso. BM!, K!).

Monochilus vieillardii Rchb.f. in Linnaea 41: 60 (1877).

Zeuxine samoensis Schltr. in Bull. Herb. Boissier. 6: 297 (1906).Type: Samoa, *Betche* 48 (holo. MEL!).

Zeuxine daenikeri Kraenzl. in Viert. Naturf. Ges. Zürich. 74: 69(1929).Type: New Caledonia, *Daeniker* 163 (holo. Z!).

Zeuxine flava sensu Jayaweera in Rev. Handb. Fl. Ceylon 2: 323 (1981); non Trimen.

Plants 30–40 cm tall. Leaves ovate, acute, 2.2–4.5 × 1.5–1.8 cm; petiole 0.5–1 cm long. Inflorescence 15–21 cm long, many-flowered; peduncle and rachis sparsely villose; bracts lanceolate, acuminate, 2–6 mm long. Flowers white, sparsely pilose on dorsal sepal; pedicel and ovary 3–6 mm long, sparsely villose. Dorsal sepal ovate, blunt, 3–4 × 2 mm. Lateral sepals obliquely ovate, obtuse, 3–4 × 2 mm. Petals linear-lanceolate, obtuse, 3–3.5 × 1–1.5 mm. Lip 3–3.5 mm long and wide; base saccate, bearing two sessile glands within; apex transversely oblong, 1.5–2 × 3.5 mm, the lobules more or less erect. Column 1.5 mm long.

DISTRIBUTION: Savai'i. Also found in New Caledonia and Fiji.

HABITAT: Terrestrial orchid found in montane forests; 500–700 m.

COLLECTIONS: *Betche* 48 (MEL); *Rechinger* 1884 p.p. (BM, W); *Vaupel* 656 (B†); *Whistler* 349 (HAW), 592 (HAW, K), 738 (BISH, HAW).

11. **SPIRANTHES**

Rich., Orch. Eur. Annot.: 20 (1817).

Small to medium-sized terrestrial herbs with fasciculate fleshy roots. Leaves in a basal rosette or in lower part of stem. Inflorescence terminal, a spiralling raceme or spike of many small flowers, often pubescent or glandular. Flowers small, usually not opening widely, white, pink or greenish. Dorsal sepal entire, adnate to the petals over the column. Lateral sepals similar but usually oblique at the base. Petals entire, membranous. Lip sessile or shortly clawed, entire or obscurely 3-lobed in apical part, hairy or glabrous on upper surface. Column rather short, lacking a foot; anther erect, dorsal; pollinia two, mealy, attached by a caudicle to a small viscidium; stigmata two, on each side under rostellum.

A genus of about 50 species, mostly in temperate North America and Central America but with a few species in Europe and temperate Asia. A single species, recorded from Samoa, extends from the Himalayas and China into S.E. Asia, the Malay Archipelago, the S.W. Pacific and northern Australia.

S. sinensis *(Pers.) Ames,* Orchid. 2: 53 (1908). Type: China, collector not traced (holo. UPS).
Neottia sinensis Pers. in Syn. 2: 511 (1807).
Spiranthes neocaledonica Schltr. in Bot. Jahrb. Syst. 39: 51 (1906). Type: New Caledonia, *Schlechter* 15594 (holo. B†).
For full synonymy see Garay & Sweet, Orchids of S. Ryukyu Islands: 63 (1974).

Plants 8–40 cm tall. Leaves mostly basal, oblong-elliptic to linear-lanceolate, up to 20 × 1 cm. Inflorescence erect, spiralled, slender, laxly many-flowered; bracts ovate-lanceolate, 5–8 mm long, glabrous. Flowers white or pink with a white lip, apparently self-pollinating; pedicel and ovary up to 6 mm long. Dorsal sepal ovate, acute, 3.2–4 × 1–1.4 mm. Lateral sepals oblong-lanceolate, acute, 2.3–4.5 × 1 mm. Petals oblong-lanceolate, rounded and erose at apex, 2.8–4 × 0.7–1 mm. Lip oblong-obovate, rounded at the apex, slightly constricted and papillose in apical third, 4–5 × 2–2.5 mm, the apical margins undulate-crispate; calli basal, subglobose, fleshy. Column 2 mm long.

DISTRIBUTION: Savai'i. Also widely distributed from mainland Asia, S.E. Asia and the Malay Archipelago to New Guinea, the Solomon Islands, Vanuatu, Niue, Australia and New Caledonia.
HABITAT: Terrestrial orchid found growing in montane ash deposits and lava flows; 370–1600 m.
COLLLECTIONS *Whistler* 2584 (BISH, HAW, K), 9387 (HAW, K), 9618 (HAW).

12. CRYPTOSTYLIS

R.Br., Prod. Fl. Nov. Holl.: 317 (1810).

Medium-sized terrestrial plants with short erect rhizomes and spreading fleshy roots. Leaves 1–3 per shoot, erect, ovate or elliptic-ovate, longly petiolate, non-articulate. Inflorescence erect, laxly few-flowered, terminal on a leafless shoot. Flowers non-resupinate, relatively large. Sepals and petals free, spreading, linear. Lip erect, much larger than the sepals and petals, entire to pandurate, lacking a spur, the disc concave. Column very short, with lateral auricles at apex; anther erect, dorsal; pollinia two, mealy, attached to a small common viscidium; rostellum obscure, more or less entire.

A genus of about 20 species distributed from Australia into the S.W. Pacific and through the Malay Archipelago into S.E. Asia. A single species in Samoa.

C. arachnites *(Blume) Hassk.*, Cat. Hort. Bot. Bog.: 48 (1844).Type: Java, *Blume* s.n. (holo. L!, iso. P!).
Zosterostylis arachnites Blume, Bijdr. Fl. Ned. Ind.: 419 (1825).
Cryptostylis stenochila Schltr. in Bot. Jahrb. Syst. 39: 49 (1906). Type: New Caledonia, *Schlechter* 15596 (holo. B†).
C. alismifolia F.Muell. in South Sci. Rec. 1: 172 (1881). Type: Samoa, 'Upolu, *Betche* s.n. (holo. MEL!).
C. vitiensis Schltr. in Repert. Spec. Nov. Regni Veg. 3: 16 (1906). Type: Fiji, *Thomson* s.n. (holo. B†).

Plants up to 40 cm tall. Leaves 1–2, elliptic to ovate-elliptic, acuminate, 10–19 × 3.5–5 cm, green often with marked longitudinal and transverse darker green venation. Inflorescence laxly 7–15-flowered; bracts lanceolate, acuminate, 7–12 mm long. Flowers green with reddish lip marked with darker red spots, apparently self-pollinating; pedicel and ovary 15–18 mm long. Sepals linear-lanceolate, 14–18 × 2.5–2.8 mm. Petals similar but 10–13 × 1.5–1.7 mm. Lip broadly lanceolate, acuminate, 15–20 × 5.5–7 mm, the disc densely puberulent. Column 2 mm long.

DISTRIBUTION: Savai'i, Ta'u, 'Upolu. Also widespread from India and tropical Asia throughout Malesia to New Guinea, and eastwards it has been found in the Solomon Islands, Vanuatu and Fiji.

HABITAT: Terrestrial orchid in rain-forests and cloud-forest; 500–1200 m.

COLLECTIONS: *Betche* s.n. (MEL); *Christophersen* 2185 (BISH, K); *Horne* 44 (K); *Whistler* 3199 (HAW), 3728 (HAW).

13. **CORYBAS**

Salisb., Parad. Londin.: t.83 (1807).

Dwarf terrestrial herbs growing from ovoid tubers. Stem very short. Leaf solitary, heart-shaped to ovate, often attractively marked on the veins, usually held horizontal to but above the substrate. Inflorescence one-flowered; bract small, linear-lanceolate. Flower solitary, subsessile, sitting on leaf or just above it. Dorsal sepal forming a hood over the column. Petals and lateral sepals often linear or rudimentary, pendent or spreading. Lip entire, strongly recurved, often with two short spurs at the base, strongly recurved in apical half. Column short, erect; pollinia four, mealy.

A genus of about 100 species widely distributed from the Himalayas and southern China, throughout S.E. Asia and the Malay Archipelago to New Guinea, Australia, New Zealand and the S.W. Pacific Islands. A single species has been recorded from Samoa.

C. betchei *(F.Muell.) Schltr.* in Repert. Spec. Nov. Regni Veg. 19: 22 (1923). Type: Samoa, *Betche* s.n.(holo. MEL!).
Corysanthes betchei F.Muell. in Wing, South Sci. Rec. 1: 171 (1881).

Plant 3–5 cm tall. Leaf cordate, apiculate, 1.8–3 × 1.5–2.6 cm. Inflorescence 1-flowered; bract linear, 4–5 mm long. Flower helmet-shaped, probably white and maroon; ovary 3–4 mm long. Dorsal sepal curved forwards, obovate, rounded at apex, 14–15 × 8 mm. Lateral sepals linear-tapering, 10–12 × 0.5 mm. Petals linear-tapering, 8–9 × 0.5 mm. Lip strongly recurved, 12 mm × 12 mm, apical margins erose; spurs short, conical, 1.5 mm long. Column 3–4 mm long.

DISTRIBUTION: 'Upolu. Known only from the type collection.
HABITAT: Terrestrial in rain-forest.
COLLECTION: *Betche* s.n. (MEL).

14. **PERISTYLUS**
Blume, Bijdr. Fl. Ned. Ind.: 404 (1825).

Terrestrial herbs growing from tubers or clustered thickened roots. Stem short or long, leafy. Leaves in a rosette or alternate along stem, entire, fleshy-membranous, not articulated. Inflorescence terminal, laxly to densely many-flowered, racemose. Flowers small, green or yellow-green, resupinate. Dorsal sepal entire, adnate to petals to form a hood over the column. Lateral sepals spreading, reflexed or erect. Petals membranous, entire. Lip connate at base with margins of the column, 3-lobed or entire, spurred at the base, rarely with a fleshy callus in the mouth of the spur. Column very short, erect; anther persistent, bilocular; pollinia two, clavate, each attached to a small viscidium; rostellum small; stigmas two, sessile, adnate to the base of the lip and auricles of the column.

A genus of about 70 species in tropical Asia, S.E. Asia, the Malay Archipelago and S.W. Pacific Islands. Two species have been recorded from Samoa.

Leaves 2 cm broad or less; flowers small; sepals 3 mm long or less; lip side
  lobes 5.5 mm long or less · · · · · · · · · · · · · · · · · **P. tradescantifolius**
Leaves 2.5 cm or more broad; flowers larger; sepals 6 mm long; lip side lobes
  7–9 mm long · **P. whistleri**

P. tradescantifolius *(Rchb.f.) Kores* in Allertonia 5(1): 12 (1989).Type: Ovalau, *Seemann* 608 (holo. W!, iso. AMES!, K!).
Habenaria tradescantifolia Rchb.f. in Seem. Fl. Vit.: 293 (1868).
H. papuana Kraenzl. in Bot. Jahrb. Syst. 18: 188 91894). Type: New Guinea, *Hellwig* 585 (holo. B†).

Fig. 2. *Peristylus tradescantifolius.* **A,** habit × ⅔; **B,** flower × 6; **C,** column, dorsal sepal and petal × 12; **D,** dorsal sepal × 8; **E,** petal × 8; **F,** lateral sepal × 8; **G,** lip × 6. All drawn from *Dennis* 2279 by Sue Wickison.

SUE
WICKISON

41

H. tradescantifolia Rchb.f. var. *pinquior* H.Fleischm. & Rech. in Denkschr. Kaiserl. Akad. Wiss., Math.-Naturwiss. Kl. 85: 250 (1910). Type: Samoa, Upolu, *Rechinger* 1802 (holo. W!).

H. cyrtostigma Schltr. in Repert. Spec. Nov. Regni Veg. 9: 83 (1910). Type: Samoa, *Rechinger* 732; Samoa, *Vaupel* 584; Samoa, *Rechinger* 1146 (all syn. B†).

Peristylus papuanus (Kraenzl.) J.J.Sm. in Nova Guinea 12: 3 (1913).

Plant 30–65 cm tall. Leaves 7–8, scattered in upper half of stem, lanceolate, long-acuminate, 8–20 × 1–2.8 cm, subtended by 5–6 sheathing cataphylls. Inflorescence 19–55 cm long, laxly many-flowered; bracts 5–14 mm long. Flowers pale green or greenish white; pedicel and ovary 10–13 mm long. Dorsal sepal ovate, obtuse, 2.3–3.2 × 1.2–2 mm. Lateral sepals obliquely oblong-elliptic, obtuse, 2.5–3 × 1–1.5 mm. Petals obliquely ovate, obtuse, 2.5–3.3 × 1.5–2 mm. Lip 3-lobed, obscurely 3-ridged at base; side lobes linear-tapering, recurving towards apex, 4.2–5.5 mm long; midlobe triangular-ligulate, fleshy, 1.5–2 mm long; spur cylindrical-fusiform, pendent or more or less parallel with ovary, 6–7.5 mm long. Column c.1 mm long.

DISTRIBUTION: Olosega, Savai'i, Ta'u, Tutuila, 'Upolu. Also found on Fiji and Tonga.

HABITAT: Terrestrial orchid found in montane forests; 200–1000 m.

COLLECTIONS: *Christophersen* 368 (BISH), 649 (BISH), 838 (BISH), 2093 (BISH), 2284 (BISH); *Graeffe* 1286 (HAW); *Long* 3120 (HAW); *Rechinger* 732 (W), 1146 (BM, W), 1802 (W); *Vaupel* 584 (B†); *Whistler* 2686 (HAW, K), 3083 (HAW, K), 3163 (BISH, HAW, K), 3745 (BISH, HAW), 3989 (HAW); *Wilkes* s.n. (W); *Yuncker* 9000a (BISH), 9150 (BISH).

P. whistleri *Cribb* **nom. nov.**

Habenaria vaupelii Schltr. in Repert. Spec. Nov. Regni Veg. 9: 84 (1910), non *H. vaupellii* Rchb.f. & Warm. (1881). Type: Samoa, *Vaupel* 465 (holo. B†).

Plant 8–150 cm or more tall. Leaves lanceolate, acuminate, 12–21 × 3–5.5 cm; petiole 1.5–3.5 cm long. Inflorescence 30–45 cm long, densely many-flowered; bracts lanceolate, 2–2.5 cm long, equalling or longer than the flowers. Flowers pale greenish yellow, small; pedicel and ovary 0.8–1.5 cm long. Dorsal sepal ovate, obtuse, 5–6 × 2.5 mm. Lateral sepals oblong-ovate, shortly mucronate, 5–6 × 2–2.5 mm. Petals entire, obliquely ovate, obtuse, 4.5–5 × 2 mm. Lip 3-lobed, with an obscure 3-lobed callus at mouth of spur; side lobes spreading, linear-tapering, 7–9 mm long; midlobe ligulate, obtuse, fleshy, 2–3 mm long; spur parallel to ovary, filiform, 8–10 mm long. Column very short, 1 mm long.

DISTRIBUTION: Savai'i, Ta'u, 'Upolu. Endemic.

HABITAT: Terrestrial orchid found in montane forest; 500–800 m.

COLLECTIONS: *Christophersen* 3291 (BISH), 3365 (BISH); *Graeffe* 1267 (HBG) & 1280 (HBG); *Vaupel* 465 (B†); *Whistler* 1620 (BISH, HAW), 2011

(BISH, HAW), 2029 (BISH, HAW), 3729 (BISH, HAW, K), 7086 (BISH, HAW); *Whitmee* 106 (K).

NOTE: Not to be confused with *Habenaria vaupellii* Rchb.f. & Warm. (1881) from Brazil.

15. HABENARIA

Willd., Sp. Plant. ed.4: 44 (1805).

Terrestrial plants growing from underground tubers. Stems erect, leafy. Leaves in a basal rosette or borne on stem, entire, membranous, not articulated. Inflorescence terminal, few to many-flowered, racemose. Flowers green, white, yellow, rarely red or orange, small to large, resupinate. Dorsal sepal entire, often adnate to the petals or the posterior petal lobes to form a hood over the column. Petals entire or bilobed. Lateral sepals spreading to reflexed, entire. Lip 3-lobed or 3-partite, rarely entire, spurred at base; side lobes spreading, sometimes further divided; midlobe usually pendent. Column erect, small to large; anther erect, persistent, bilocular; pollinia two, clavate, each attached to a small viscidium; rostellum 3-lobed, more or less triangular; stigmas two, stalked.

A genus of between 600 and 800 species, pantropical in distribution, but extending into warm temperate regions.

Largest leaves more than 14 cm long; spur 14 mm or more long
· **H. monogyne**
Largest leaves less than 12.5 cm long; spur less than 10 mm long
· **H. samoensis**

H. monogyne *Schltr.* in Repert. Spec. Nov. Regni Veg. 3: 45 (1906). Type: Samoa, *Betche* s.n. (holo. B†).
H. supervacanea sensu H.Fleischm. & Rech in Denkschr. Kaiserl. Akad. Wiss., Math.-Naturwiss. Kl. 85: 250 (1910), non Rchb.f.

Plant up to 60 cm tall. Leaves lanceolate, acuminate, up to 17×2.7 cm. Flowers green. Dorsal sepal ovate, acuminate, 9 mm long. Lateral sepals similar. Petals bipartite; posterior lobe linear-lanceolate, 9 mm long; anterior lobe half length of posterior lobe. Lip 3-lobed; midlobe linear, 10 mm long; side lobes up to 20 mm long; spur 14 mm long.

DISTRIBUTION: Savai'i, Tutuila, 'Upolu. Endemic to Samoa.
HABITAT: Uncommon in lowland to montane forest; 300–900 m.
COLLECTIONS: *Betche* s.n. (B†); *Christophersen* 2073 (BISH); *Rechinger* 952 (W), 1392 (W); *Setchell* 547 (UC); *Vaupel* 69 (B†).

NOTE: Allied to *H. supervacanea* Rchb.f. (1868) but with much smaller flowers.

H. samoensis *F. Muell. & Kraenzl.* in Bot. Jahrb. Syst. 17: 487 (1893). Type: Samoa, *Betche* 112 (holo. B†).

H. superflua sensu H.Fleischm. & Rech. in Denkschr. Kaiserl. Akad. Wiss., Math.- Naturwiss. Kl. 85: 250 (1910), non Rchb.f. (1868).

H. dolichostachya sensu Kraenzl. in Bot. Jahrb. Syst. 25: 558–708 (1898), non Thwaites (1864).

Plants to 60 cm tall. Leaves in central part of stem, elliptic-lanceolate to obovate, acuminate, 6–11 × 1.5–3.3 cm. Inflorescence laxly to subdensely many-flowered, 25–33 cm tall; bracts lanceolate, acuminate, 7–14 mm long. Flowers green, 1.5–2 cm tall; pedicel and ovary 7–10 mm long. Dorsal sepal ovate-elliptic, aristate, 6–7 × 2.2–2.5 mm. Lateral sepals obliquely ovate, aristate, 7–8 × 2.5 mm. Petals bipartite; posterior lobe broadly linear-lanceolate, acuminate, 4–5.5 × 1–1.2 mm; anterior lobe falcate, much shorter than posterior lobe, linear, acuminate, 3–3.5 × 0.3–0.4 mm. Lip 3-lobed; side lobes linear-tapering. 4–5.5 mm long; midlobe a little longer than side lobes, linear, acute, 5–6 mm long; spur more or less equal in length to ovary, 8–10 mm long, filiform, slightly inflated at apex.

DISTRIBUTION: Savai'i, Ta'u. Endemic to Samoa.

HABITAT: Terrestrial orchid, uncommon in lowland and cloud-forest; 200–1550 m.

COLLECTIONS: *Betche* 112 (B†); *Rechinger* 81 (W); *Reinecke* 260 (B†), 260a (B†), 260b (B†), 260c (B†); Whistler 1984 (BISH, HAW), 2482 (BISH, HAW, K), 2626 (BISH, HAW), 9607 (HAW).

NOTE: Similar to *H. superflua* Rchb.f. (1868) but much smaller in all its floral parts and with shorter and broader leaves.

16. NERVILIA

Commerson ex Gaudich. in Freycinet, Voy. Uranie et Physicienne, Bot.: 421 (1829).

Small to medium-sized terrestrial herbs growing from underground tubers. Leaf solitary, usually produced after the inflorescence has withered, ovate, lanceolate elliptic or heart-shaped, erect or borne parallel to the substrate, either appressed to the substrate or above it, glabrous to hairy, green or marked with purple on upper side, green or purple beneath. Inflorescence lateral, 1–several-flowered, erect, racemose. Flowers short-lived, resupinate. Sepals subsimilar, linear-lanceolate. Petals similar but shorter and membranous. Lip more or less embracing the column, entire to 3-lobed, bearing a lamellate or hairy callus, rarely spurred at base. Column clavate, elongate; anther incumbent, 2-celled; pollinia two, granulose, lacking caudicles; stigma below rostellum.

A genus of about 80 species in Africa, Madagascar, tropical Asia, the Malay Archipelago, Australia and the S.W. Pacific Islands. Two species reported from Samoa.

Flowers 5 or more per inflorescence; sepals 2.5 cm long or less; lip 3-lobed at the apex, pilose on the upper surface · · · · · · · · · · · · · · · **N. aragoana**
Flowers 2 per inflorescence; sepals 3 cm long; lip entire, lacking a callus or pubescence · **N. grandiflora**

N. aragoana *Gaudich.* in Freycinet, Voy. Uranie et Physicienne, Bot.: 422 (1829). Type: Mariana Islands, *Gaudichaud* s.n. (holo. P!)
Pogonia flabelliformis Lindl., Gen. Spec. Orch. Pl.: 415 (1840), *nom. illeg.* Type: Nepal, *Wallich* 7400A (holo. K!)
P. nervilia Blume in Mus. Bot. Lugduno-Bat. 1: 32 (1849), *nom. illeg.*, based on the types of both above.
Gastrodia sp. sensu Kraenzl. in Bot. Jahrb. Syst. 25: 599 (1898).

Leaf erect, plicate, heart-shaped, acute, 12–15 × 14–18 cm, often marked with dark maroon on upper surface; petiole 15–30 cm long. Inflorescence laxly 5–15-flowered, up to 45 cm tall; bracts linear to linear-lanceolate, 15–25 mm long. Flowers pendent or nodding, probably self-pollinating, greenish yellow with a white lip marked with rose or violet veins; pedicel and ovary 1–1.5 cm long. Sepals and petals linear-lanceolate, acute, 2–2.5 cm long. Lip 3-lobed near the apex, 2–2.4 cm long; side lobes small, erect, triangular; midlobe subovate, acute to obtuse, with undulate margins; callus puberulent in middle. Column clavate, 7 mm long.

DISTRIBUTION: Ofu, Savai'i, Tutuila, 'Upolu. Also widely distributed from Asia and the Malay Archipelago to the Mariana Islands, New Guinea, the Solomon Islands, Vanuatu, New Caledonia, the Horne Islands, Society Islands, Fiji, Niue Island and Australia.
HABITAT: Terrestrial orchid found in rain-forest; 10–500 m.
COLLECTIONS: *Christophersen* 2878 (BISH), 3406 (BISH); *Rechinger* 1948 (BM, W), 5271 (W); *Reinecke* 140 (B†), 601 (B†); *Vaupel* 508 (B†); *Whistler* 3024 (HAW), 3478 (HAW), 3776 (HAW), 7008 (HAW), 8084 (HAW).

N. grandiflora *Schltr.* in Repert. Spec. Nov. Regni Veg. 9: 85 (1910). Type: Samoa, *Vaupel* 590 (holo. B†), non *N. grandiflora* Schltr. in Ann. Transv. Mus. 10: 241 (1924).

Plant about 35 cm tall. Leaf subreniform-cordate, shortly apiculate, 7–8 × 10–11 cm, glabrous on both surfaces; petiole about 12 cm long. Inflorescence 2-flowered. Flowers erect-spreading; pedicel and ovary about 13 mm long. Sepals lanceolate, acuminate, 30 × 4.5 mm; the laterals slightly oblique. Petals similar to sepals, about 24 mm long. Lip entire, elliptic, subtruncate or very obtuse at apex, 18 × 8 mm, slightly thickened longitudinally in the centre. Column 10 mm long, glabrous.

DISTRIBUTION: Savai'i. Endemic.
HABITAT: Terrestrial orchid found in rain-forest.

COLLECTIONS: Known only from the type.

NOTE: This is close to *N. platychila* Schltr.(1906) from New Caledonia and Fiji but apparently differs in having a glabrous leaf. The destruction of the type collection and the lack of other collections from Samoa make further comment difficult. Nervilias are easily overlooked as the flowers appear before the leaf develops. It would be worthwhile searching for this orchid again on Savai'i.

17. DIDYMOPLEXIS

Griffith in Calcutta J. Nat. Hist. 4: 383 (1843).

Small saprophytic plants growing from a tuberous rhizome. Stems erect, glabrous, bearing a few cataphylls, lacking chlorophyll and leaves. Inflorescence few-flowered, racemose. Flowers resupinate, small, lacking a spur. Dorsal sepal adnate to petals and forming a hood over the column. Lateral sepals free or fused to each other for at least part of length. Lip 3-lobed, attached to column-foot; callus of rows of papillae. Column free, elongate, with a short foot; anther declinate; pollinia two, sectile, lacking caudicles; rostellum short. Fruits with a stalk that elongates rapidly after pollination.

A small genus of about 20 species in tropical Africa, Madagascar, tropical Asia across to the Ryukyu Islands, the Malay Archipelago, Northern Australia and the S.W. Pacific. A single species reported from Samoa.

D. micradenia *(Rchb.f.) Hemsl.* in J. Linn. Soc. Bot. 20: 311 (1883). Type: Ovalau, *Seemann* 610 (holo. W!, iso. K!).
Epiphanes micradenia Rchb.f. in Seem., Fl. Vit.: 295 (1868).
Leucorchis micradenia (Rchb.f.) Benth. & Hook.f. ex Drake in Ill. Fl. Ins. Mar. Pac.: 313 (1892).
Didymoplexis neocaledonica Schltr. in Bot. Jahrb. Syst. 39: 50 (1906).Type: New Caledonia, *Schlechter* 15748 (holo. B†).
D. minor J.J.Sm. in Bull. Inst. Bogor. 7: 1 (1900).Type: Java, *J.J.Smith* 74 (holo. L!).
D. minor subsp. *samoensis* H.Fleischm. & Rech. in Denkschr. Kaiserl. Akad. Wiss., Math.-Naturwiss. Kl. 85: 251 (1910). Type: Samoa, *Rechinger* 1641 (holo. W!).
D. samoensis (H.Fleischm. & Rech.) Schltr. in Repert. Spec. Nov. Regni Veg. 9: 85 (1910).
D. pallens sensu Sykes in New Zealand Dept. Sci. Indust. Res. Bull. 200: 258 (1970), non Griff.

Fig. 3. *Didymoplexis micradenia.* **A**, habit × ⅔; **B**, flower × 3; **C**, lip × 6. *Stereosandra javanica.* **D**, habit × ⅔; **E**, flower × 3; **F**, lip × 6. **A–C** drawn from *Cribb & Wheatley* 3; **D–F** from *Mitchell* 35. All drawn by Sue Wickison.

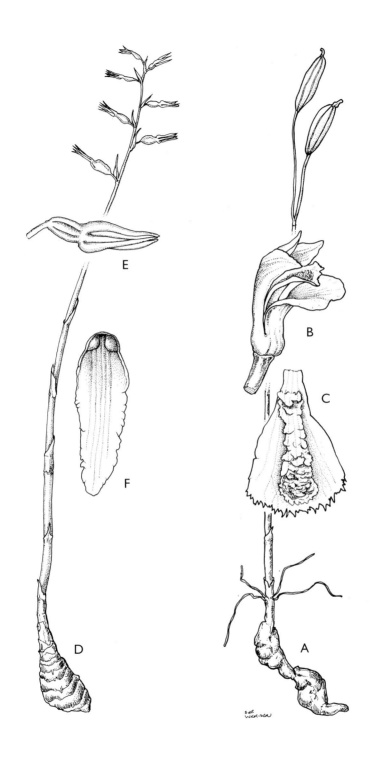

E

B

C

F

D

A

WICKISON

Plant 7–10 cm tall, elongated in fruit to up to 25 cm tall. Inflorescence erect, slender, up to 10 cm tall, pinkish brown; peduncle bearing 2–3 short sheathing cataphylls; bracts ovate-acuminate, 1–1.5 mm long. Flowers small, dull flesh-brown with a whitish lip; peduncle elongating rapidly after fertilization, 5–20 cm long. Dorsal sepal oblong-ovate, blunt, adnate to petals forming a hood over column, 5.5–6.6 × 2.5–3.5 mm. Lateral sepals almost fused, oblong to oblong-lanceolate, obtuse, 5–6 × 1.5–2 mm. Lip oblong-cuneiform, weakly 3-lobed at apex, 5.5–6.5 × 2.5–3 mm; callus of 3 ridges of papillae. Column 4.5–5.5 mm long; foot 0.3–0.5 mm long.

DISTRIBUTION: Ofu, Olosega, Savai'i, Ta'u, Tutuila, 'Upolu. Also in Vanuatu, New Caledonia, Fiji, Tonga and Niue.

HABITAT: Found in coastal and lowland forests; near sea level to 450 m.

COLLECTIONS: *Rechinger* 1641 (W); *Vaupel* 600 (B†); *Whistler* 1026 (HAW), 2847 (HAW), 2924 (HAW), 3023 (HAW), 3069 (BISH, HAW), 3133 (BISH, HAW, K), 3656 (HAW), 3988 HAW), 4024 (HAW), 4167 (HAW), 8104 (HAW), 8325 (HAW), 9330 (HAW), 9345 (HAW); *Wilkes* s.n. (W).

18. **STEREOSANDRA**

Blume in Mus. Bot. Lugd.-Bat. 2: 179 (1856).

Saprophytic, terrestrial, leafless herbs. Rhizomes tuberous, several-noded. Inflorescence erect, unbranched, laxly few- to many-flowered; peduncle terete, bearing several well-spaced sterile, lanceolate, sheathing bracts; fertile bracts persistent, lanceolate. Flowers spreading to subpendent, not opening widely, out-crossing or self-pollinating. Sepals and petals free, subsimilar. Lip entire, bearing two basal calli, lacking a spur. Column short; anther erect on a broad filament, rising from the back of the coloumn; pollinia 2.

A small genus of about five species from S.E. Asia, New Guinea, the Solomon Islands and Samoa, whence a single species is recorded.

S. javanica *Blume* in Mus. Bot. Lugd. Bat. 2: 176 (1856). Type: Java, *Blume* s.n. (holo. L!).

Plants 20–25 cm tall, arising from a fleshy, erect or suberect, cylindrical or ellipsoidal rhizome, 2–3.5 × 0.4–1.5 cm. Inflorescence very short-lived, laxly few-flowered; pedicel, rachis and bracts pale buff; rachis 3–5 cm long; bracts lanceolate, acuminate, 3–6 mm long. Flowers white with the tips of the sepals, petals and lip tinged with purple; pedicel and ovary 2–3 mm long, purple-striped. Sepals lanceolate, acute, 7–9 × 2 mm. Petals lanceolate, acute, 7–8 × 1.5 mm. Lip ovate or cymbiform, obtuse, 8–9 × 2.5 mm, bearing two fleshy, round calli at the base. Column 3–4 mm long. Fruit ellipsoidal, 7 mm long, perianth persistent.

DISTRIBUTION: Tutuila. Also in the Malay Peninsula and Archipelago, New Guinea and the Solomon Islands. This orchid only appears above the ground when flowering and fruiting, a process that takes only a few days to complete. It is certainly under-represented in collections.

HABITAT: In deep leaf litter in dense shade in forest; up to 600 m.

COLLECTION: *Sykes* 36 (CHR, apparently lost).

19. **CALANTHE**

R.Br. in Ker-Gawler in Bot. Reg. 7: sub t.573 (1821).

Medium-sized to large terrestrial or rarely epiphytic plants with a short to elongate rhizome. Erect stems pseudobulbous or obscurely so, several-noded, leafy along length. Leaves pleated, not articulated or rarely articulated, often quite large. Inflorescence lateral, erect, many-flowered, racemose. Flowers resupinate, usually showy, turning blue-black when damaged and with age; bracts persistent or rarely caducous. Sepals free, spreading. Petals smaller than sepals, free, spreading. Lip entire or more commonly 3- or 4-lobed, connate at base to column, spurred at base; callus papillate, ridged or verrucose, at base of lip; spur usually long, rarely short or absent. Column short, fleshy, connate with base of lip; anther decumbent; pollinia 8, clavate, waxy.

A genus of about 260 species widely distributed in Africa, Madagascar, tropical and east temperate Asia, Japan, Taiwan, S.E. Asia, the Malay Archipelago, W. and S.W. Pacific Islands, Australia and a single species in the tropical Americas. Four species have been recorded from Samoa.

1. Flowers non-resupinate · **C. alta**
   Flowers resupinate ·2
2. Flowers yellow · **C. ventilabrum**
   Flowers white, sometimes with a yellow callus · · · · · · · · · · · · · · · · · · ·3
3. Spur slender, straight or incurved-arcuate, puberulent; lip deeply 4-lobed,
   with a basal verrucose callus · **C. triplicata**
   Spur subsigmoid, glabrous; lip obscurely 3-lobed, ecallose · **C. hololeuca**

C. alta *Rchb.f.* in Otia Bot. Hamburg.: 53 (1878). Type: Samoa, U.S. Expl. Exped., *Wilkes* s.n. (holo. AMES!, iso. W!).
C. lutescens H.Fleischm. & Rech. in Denkschr. Kaiserl. Akad. Wiss., Math.-Naturwiss. Kl. 85: 257 (1910). Type: Samoa, *Rechinger* 1826 (holo. W!).
C. anocentrum Schltr. in Repert. Spec. Nov. Regni Veg. 9: 98 (1910). Type: Samoa, *Vaupel* 89 (holo. B†; iso. AMES!, BISH!, K!, US, W!).

Plants up to 120 cm tall. Leaves elliptic to elliptic-oblanceolate, acuminate, 35–55 × 7–13 cm; petiole 15–25 cm long. Inflorescence 60–80 cm tall, subdensely many-flowered; peduncle twice as long as the rachis; bracts lanceolate, 10–15 mm long, puberulent. Flowers white with yellow at apex of column; pedicel and ovary 10–18 mm long. Sepals elliptic-ovate, obtuse, 8–10 × 6–7 mm. Petals elliptic-spathulate, obtuse, 6–9 × 3–4 mm. Lip deeply 3-lobed, 7–11 × 8–20 mm; side lobes oblong-elliptic, rounded at apex; midlobe oblong, deeply retuse or bifid at apex; callus at apex of column transverse, multi-tuberculate; spur cylindric, 8–12 mm long, laxly puberulent at apex. Column 3 mm long.

DISTRIBUTION: Savai'i and 'Upolu. Also from Fiji.
HABITAT: Terrestrial orchid in rain-forest and cloud-forest; 600–800 m.
COLLECTIONS: *Rechinger* 1826 (W); *Vaupel* 89 (AMES, BISH, K, US, W); *Whistler* 350 (HAW), 1663 (BISH, HAW), 6829 (BISH, HAW); *Wilkes* s.n. (AMES, W).

C. hololeuca *Rchb.f.* in Seem., Fl. Vit.: 298 (1868). Type: Viti Levu, *Seemann* 607 (holo. K!, part of holo. W!).
C. vaupeliana Kraenzl. in Notizbl. Bot. Gart. Berlin-Dahlem 5: 111 (1909). Type: Samoa, *Vaupel* 358 (holo. B †, iso. AMES!, K!).
C. neocaledonica Rendle in J. Linn. Soc. Bot. 45: 251 (1921). Types: New Caledonia, *Compton* 1409 & 1609 (syn. BM!).
C. clavata sensu Kraenzl. in Bot. Jahrb. Syst. 25: 603 (1898) & sensu H.Fleischm. & Rech. in Denkschr. Kaiserl. Akad. Wiss., Math.-Naturwiss. Kl. 85: 257 (1910), non Lindl.
C. sp. sensu Yuncker in Bernice P. Bishop Mus. Bull. 1184: 32 (1945).

Plants 50–90 cm tall. Leaves lanceolate, acute or acuminate, 30–50 × 3–6.8 cm; petiole 11–20 cm long. Inflorescence 27–65 cm long; subdensely many-flowered; peduncle terete; rachis 6–15 cm long; bracts ovate-lanceolate, acute, deciduous, 15–28 mm long. Flowers white; pedicel and ovary 10–20 mm long. Sepals elliptic to oblong-ovate, acuminate or shortly apiculate, 11–14 × 4–6 mm. Petals elliptic to elliptic-obovate, acuminate, 10–13 × 4.5–8 mm. Lip 3-lobed, 6–7 × 4–5 mm, lacking a basal callus; side lobes upcurved, oblong, small; midlobe oblong-cuneate or oblong, truncate; spur cylindric-slightly clavate, 12–14 mm long, slightly sigmoid at apex. Column 3 mm long.

DISTRIBUTION: Savai'i, Ta'u, Tutuila, 'Upolu. Also from the Santa Cruz Islands, Vanuatu, New Caledonia, Fiji, Tonga and the Horne Islands.
HABITAT: Terrestrial orchid common in montane forest; 300 to 1200 m.
COLLECTIONS: *Christophersen* 1020 (BISH, P), 2086 (AMES, BISH, P), 3501 (BISH), 3579 (BISH); *Garber* 783 (BISH), 784 (BISH), 785 (BISH); *Rechinger* 1695 (W), 1979 (W); *Reinecke* 455 (B†); *Setchell* 376, 546; *Vaupel* 358 (AMES, K); *Whistler* 1406 (HAW), 1898 (HAW), 2980 (HAW), 3159 (BISH, HAW, K), 7824 (HAW), 8009 (HAW), 9183 (HAW); *Wilkes* s.n. (W); *Yuncker* 9248 (BISH).

C. triplicata *(Willemet) Ames* in Philipp. J. Sci., Bot. 2: 326 (1907).
Orchis triplicata Willemet in Ann. Bot. Usteri 18: 52 (1796). Type: Amboina,
Rumphius s.n. (holo. L!).
Limodorum veratrifolium Willd., Sp. Plant. ed.4, 4: 122 (1804), nom. illeg.
Calanthe veratrifolia (Willd.) R. Br. in Bot. Reg. 7: sub t. 573 (1821); Ker-Gawl.,
op. cit. 9: t. 720 (1823).
C. angraeciflora Rchb.f. in Linnaea 41: 75 (1877). Type: New Caledonia,
Deplanche 114 (holo. P!).
C. nephroglossa Schltr. in Repert. Spec. Nov. Regni Veg. 9: 99 (1911). Type:
Samoa, *Vaupel* 413 (holo. B†, iso. K?), **synon. nov.**
C. furcata sensu Yuncker in Bernice P. Bishop Mus. Bull. 184: 32 (1945).
C. triplicata var. *angraeciflora* (Rchb.f) N.Hallé in Fl. Nouv. Caled. 8: 230
(1977). Type: As for *C. angraeciflora*
For a full synonymy see Garay and Sweet, Orchids S. Ryukyu Islands: 119
(1974).

Plants 60–125 cm tall. Leaves elliptic-lanceolate, acuminate, 40–75 × 4–9
cm; petiole 12–25 cm long. Inflorescence 60–125 cm long, laxly to subdensely
many-flowered; peduncle pubescent, terete; rachis a quarter to half length of
peduncle; bracts lanceolate, acuminate. Flowers white with a whitish or yellow
callus; pedicel and ovary 2–3 cm long. Sepals elliptic-obovate, abruptly
acuminate, 12–20 × 6–9 mm. Petals elliptic-oblanceolate or oblanceolate,
acute, 10–18 × 4–7 mm. Lip 4-lobed, 13–20 × 11–14 mm; side lobes obliquely
oblong-elliptic, rounded at apex; midlobes falcate, linear-oblong, spreading;
callus of three short verrucose ridges at apex of column; spur filiform, 15–30
mm long, arcuate-dependent. Column 4–5 mm long.

DISTRIBUTION: Ofu, Olosega, Savai'i, Ta'u, Tutuila, 'Upolu. Widely
distributed from Madagascar, Asia and S.E. Asia to the Malay Archipelago, New
Guinea, Solomon Islands, Vanuatu, New Caledonia, Fiji, the Marquesas, Lord
Howe Island and Australia.
HABITAT: Terrestrial orchid in forest; sea level–1600 m.
COLLECTIONS: *Bryan* 86 (BISH), 169 (BISH); *Christophersen* 964 (BISH),
1126 (BISH), 1827 (BISH), 2128 (BISH), 2268 (BISH), 2566 (BISH), 2672
(BISH), 2851 (BISH), 2862 (BISH, P); *Eames* 208 (BISH); *Garber* 1036 (BISH),
1077 (BISH); *Graeffe* 1287 (HAW, HBG); *Kennedy* 4000 (BISH, P); *Horne* s.n.
(K); *Long* 3050 (HAW); *McKee* 3018 (BISH); *Mitchell* 561 (BISH); *Powell* 271
(K); *Rechinger* 366 p.p.(W); *Reinecke* 185 (B†), 293 (B†); *Setchell* 408 (UC);
Vaupel 409 (B†), 413 (B†, K); *Whistler* 888 (HAW), 1042 (HAW), 1225 (HAW),
1600 (BISH, HAW), 1771 (BISH), 1898a (HAW), 1995 (HAW), 2509 (HAW),
3025 (BISH, HAW), 4772 (HAW), 7017 (HAW), 8215 (HAW); *Wisner* 142
(BISH); *Yuncker* 9149 (BISH).

C. ventilabrum *Rchb.f.* in Seem., Fl. Vit.: 298 (1868). Type: Fiji, Taveuni,
Seemann 606 (holo. W, apparently lost; iso. K!; drawing of holo. at AMES!).
C. langei F.Muell. in South. Sci. Rec. n.s.: 1 (1885). Type: New Caledonia,
Layard s.n. (holo. MEL!).

C. bigibba Schltr. in Repert. Spec. Nov. Regni Veg. 9: 99 (1911). Type: Samoa, *Vaupel* 359 (holo. B †; iso. AMES!).

Plants up to 60 cm tall. Leaves 4–6, lanceolate, acuminate, 25–60 × 4–10 cm; petiole 15–30 cm long. Inflorescence 35–65 cm long, densely many-flowered; peduncle 3 times as long as rachis; bracts deciduous, lanceolate, acuminate, 15–30 mm long. Flowers golden yellow to bright yellow, probably self-pollinating in Samoa; pedicel and ovary 1.5–2.5 cm long. Sepals ovate, acute, 8–13 × 4.5–5.5 mm. Petals elliptic, acute or subacute, 7–10.5 × 5.5–6.5 mm. Lip obscurely 3-lobed, 5–7.5 × 3–3.5 mm, ecallose; side lobes semi-elliptic, 1–1.5 mm wide; midlobe obovate, shortly apiculate, 3.5–5.5 mm long; spur subsigmoid, saccate-clavate, 4.5–7.5 mm long. Column 2–2.5 mm long.

DISTRIBUTION: Savai'i. Also in New Guinea, Bougainville, the Solomon Islands, Vanuatu, New Caledonia and Fiji.
HABITAT: Terrestrial orchid in cloud-forest; 1200–1700 m.
COLLECTIONS: *Christophersen* 813 (BISH, K), 2137 (BISH), 2216 (BISH); *Cox* 267 (BISH); *Powell* 253 (BISH); *Vaupel* 359 (AMES); *Whistler* 2479 (HAW), 2558 (HAW).

20. **PHAIUS**

Lour., Fl. Cochinchin.: 517 (1790).

Large terrestrial plants with short stout rhizomes. Stems pseudobulbous, or elongate and cane-like, leafy. Leaves pleated, large, not articulated. Inflorescence few to many-flowered, racemose, axillary. Flowers usually large and often showy, resupinate, turning blue-black when damaged or with age; bracts persistent. Sepals and petals free, usually spreading, similar in shape and size. Lip shortly connate to column at the base, 3-lobed or entire, spurred at base, usually enclosing the column at the base; callus carinate. Column elongate, fleshy, lacking a foot or with a very short one; anther terminal; pollinia 8 in two groups of four, clavate, waxy.

A genus of about 40 species in Africa, Madagascar, tropical Asia, the Malay Archipelago, N.E. Australia eastwards into the S.W. Pacific across to the Society Islands. Introduced into Hawaii. Three species reported from Samoa.

Fig. 4. *Calanthe ventilabrum.* **A**, habit × ⅔; **B**, inflorescence × 3; **C**, flower × 3; **D**, lip, spur and column × 4; **E**, dorsal sepal × 4; **F**, petal × 4; **G**, lateral sepal × 4; **H**, lip × 6; **J**, column × 8; **K**, dorsal and ventral view of anther cap × 8; **L**, pollinia × 8. *C. hololeuca.* **M**, flower × 3; **N**, spur, lip and column × 3; **O**, lip × 4. **A–L** drawn from *Wickison* 36: **M–O** from *Wickison* 124. All drawn by Sue Wickison.

1. Flowers white, sometimes with a creamy lip; stem elongate; lip lacking a spur; bracts caducous · **P. terrestris**
   Flowers yellow or brownish with red or purple on the lip; stem short, covered by leaf bases; lip spurred at base; bracts persistent · · · · · · · · · **2**
2. Flowers yellow with a red margin to the lip; lip with undulate blunt apex; leaves often spotted with yellow · **P. flavus**
   Flowers with chestnut brown sepals and petals and a whitish lip marked with purple; lip apex obtuse or apiculate; leaves green   **P. tankervilleae**

P. flavus *(Blume) Lindl.*, Gen. Sp. Orch. Pl.: 128 (1831). Type: Java, *Blume* (holo. L!).
Limodorum flavum Blume, Bijdr. Fl. Ned. Ind.: 375 (1825).
Phaius maculatus Lindl., Gen. Spec. Orch. Pl.: 127 (1831). Type: Nepal, *Wallich* s.n. (holo. K!).
P. blumei sensu Kraenzl. in Bot. Jahrb. Syst. 25: 602 (1898); non Lindl.

A large terrestrial plant up to 90 cm tall. Pseudobulbs conical, 3–8-leaved at the apex. Leaves lanceolate to oblong-lanceolate, acuminate, 20–48 × 3–11 cm, green sometimes spotted with yellow. Inflorescence erect, 50–90 cm tall, laxly several-flowered; bracts persistent, 1.5–2.8 cm long. Flowers yellow, lip paler at base and marked with brownish red at the apex; pedicel and ovary 1.4–3.5 cm long. Sepals oblong, blunt, 2.8–4 × 1–1.3 cm ; lateral sepals falcate. Petals oblong, 2.7–3.5 cm long. Lip connate for basal 4 mm with the column, rhombic-orbicular, somewhat 3-lobed in front, 3.5–4 × 3–3.5 cm, apical margins undulate-crispate; spur conical, 4–6 mm long. Column hairy on ventral surface, 1.6–2 cm long.

DISTRIBUTION: Savai'i. Nepal to Peninsular Malaysia, the Malay Archipelago, Vanuatu and New Caledonia.
HABITAT: Terrestrial orchid found in cloud-forests; 1500–1550 m.
COLLECTIONS: *Reinecke* 456 (B†); *Whistler* 2508 (HAW, K), 9599 (HAW).

P. tankervilleae *(Banks) Blume* in Ann. Mus. Bot. Lugduno-Batavia 2: 177 (1856); as *P.tankervilii.* Type: China, cult. Kew (holo. BM!).
For full synonymy see Kores in A.C. Smith, Fl. Vit. Nov. 5: 474 (1991).

A large terrestrial herb up to 2 m tall. Pseudobulbs conical to ovoid, 2.5–6 × 2–5 cm. Leaves narrowly elliptic to elliptic-lanceolate, acuminate, up to 120 × 4–20 cm; petiole up to 25 cm long. Inflorescence erect, laxly 10–20-

Fig. 5. *Phaius terrestris.* **A**, habit × ⅑; **B**, flower × ⅔; **C**, dorsal sepal × 1; **D**, petal × 1.5; **E**, lateral sepal × 1.5; **F,G**, lip side view and flattened × 1.5; **H**, column × 1.5; **J**, pollinia × 3; **K**, anther cap two views × 3; **L**, flower from front × ⅔. **A–K** drawn from *Wickison* 92, **L** from a colour slide. All by Sue Wickison.

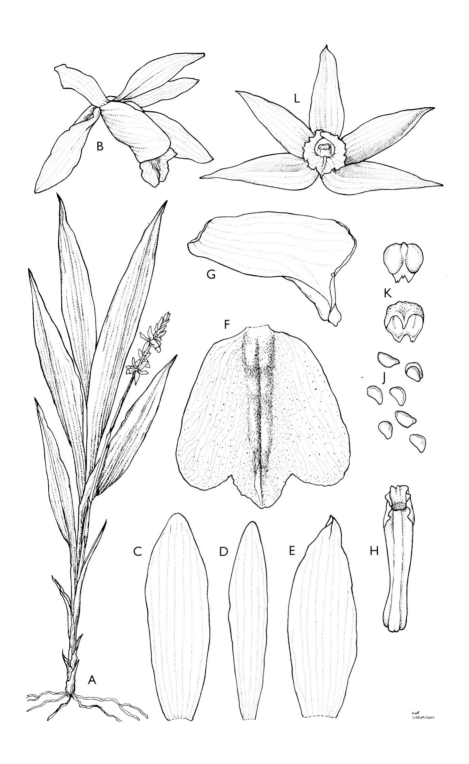

55

flowered, 60–200 cm tall; bracts persistent, lanceolate, 3–5 cm long. Flowers large, showy, the sepals and petals white to pale purple on outer side, yellow to brown within; lip white or yellowish in throat, heavily marked with purple especially towards the apex; pedicel and ovary 2.5–5 cm long. Sepals lanceolate to oblanceolate, acuminate, 4.5–6.5 × 0.7–1.5 cm. Petals similar. Lip trumpet-shaped, broadly obovate and obscurely 3-lobed when flattened, apiculate, 4–5.5 cm long; spur slightly decurved, 0.6–0.8 cm long. Column slightly clavate, 1.5–2 cm long.

DISTRIBUTION: Ofu, Olosega, Savai'i, Ta'u, Upolu. Also widely distributed from Asia, S.E. Asia and the Malay Archipelago to New Guinea, New Caledonia, the Horne Islands, Fiji and Australia. Probably a recent arrival in Samoa and elsewhere in the Pacific Islands.

HABITAT: Terrestrial orchid found in montane forest; 200–700 m.

COLLECTIONS: *Christophersen* 2085 (BISH); *Whistler* 1955 (BISH, HAW), 3071 (BISH, HAW), 3185 (HAW), 7807 (HAW), 9511 (HAW); *Yuncker* 9261 (BISH).

P. terrestris *(L.) Ormerod* in Austral. Orchid Rev. 59: 14 (1994). Type: Based on Rumphius, Herb. Amboin. 6: t. 52, f.1 (1750).
Epidendrum terrestre L., Syst. Veg. ed.10, 2: 1246 (1759).
Phaius amboinensis Blume, Mus. Bot. Lugd.-Batavia 2: 180 (1852). Type: Ambon, *Zippelius* s.n. (lecto. L!), selected here.
P. graeffei Rchb.f. in Seem., Fl. Vit.: 299 (1868), (repr. Xenia Orch. 3: 30 (1881)). Type: Samoa, *Graeffe* s.n. (holo. W!).

A large terrestrial plant up to 1 m tall. Stems elongate, non-pseudobulbous, up to 30 cm long, leafy. Leaves elliptic-lanceolate, acuminate, up to 65 × 4–9 cm, petiolate. Inflorescence erect, axillary, 30–70 cm long, laxly 5–15-flowered; bracts obovate, 2.5–3.5 cm long, caducous. Flowers showy, white with a pale yellowish lip, turning blue when damaged, sometimes cleistogamous; pedicel and ovary 1.8–3 cm long. Sepals oblong-obovate, subacute, 2.8–3.5 × 1–1.3 cm. Petals oblanceolate, slightly falcate, rounded at apex, 2.8–3.3 × 0.7–0.8 cm. Lip embracing column, flabellate when flattened, 3-lobed at apex, 2.5–3 cm long and wide, lacking a spur, mealy tomentose on the disc. Column clavate, 2–2.5 cm long.

DISTRIBUTION: Olosega, Savai'i, Tutuila, 'Upolu. Also in New Guinea, Bougainville, the Solomon Islands, Vanuatu, the Cook Islands and Fiji.

HABITAT: Terrestrial orchid found in rain-forest; 50–800 m.

COLLECTIONS: *Christophersen* 49 (BISH); *Cox* 145 (BISH), 183 (BISH), 305 (BISH); *Graeffe* s.n. (W); *Rechinger* 25 (BM, W), 366 p.p. (W), 404 (W); *Whistler* 1570 (HAW), 1900 (HAW), 1966 (BISH, HAW), 2559 (BISH, HAW), 2687 (BISH, HAW), 3026 (BISH, HAW), 3072 (HAW), 3132 (HAW), 5707 (HAW), 6892 (HAW), 6929 (HAW), 8219 (HAW), 8525 (HAW), 8774 (HAW).

21. **SPATHOGLOTTIS**

Blume, Bijdr. Fl. Ned. Ind. 1,8: 400 (1825).

Medium-sized to large terrestrial herbs. Stems clustered, pseudobulbous, hidden by leaf bases. Leaves several, suberect to spreading, pleated. Inflorescence basal, axillary, erect, laxly to densely several–many-flowered; bracts persistent. Flowers showy, white, yellow, pink or purple, often with a yellow callus on lip. Sepals subsimilar, free, spreading widely. Petals similar, often slightly smaller than the sepals. Lip 3-lobed at base, lacking a spur, with a callus; side lobes upcurved-erect, smaller than midlobe; midlobe linear, spathulate or obovate; callus between lateral lobes, entire or bilobed, glabrous or hairy. Column clavate, lacking a foot; pollinia 8, pear-shaped.

A genus of about 40 species widespread in tropical and subtropical Asia, S.E. Asia, the Malay Archipelago, New Guinea, the Philippines, N.E. Australia and the western Pacific islands. A single species *Spathoglottis plicata* has been found in Samoa. It is found as an escape in many parts of the tropics including Hawaii and Kenya.

S. plicata *Blume*, Bijdr. Fl. Ned. Ind.: 401 (1825). Type: Java, *Blume* s.n. (holo. L!, iso. P!).

S. vieillardii Rchb.f. in Linnaea 41: 85 (1877). Type: New Caledonia, *Vieillard* 1302 (holo. P!).

Bletia angustifolia Gaudich. in Freycinet, Voy. Uranie et Physicienne: 421 (1829). Type: Moluccas, *Freycinet* s.n. (holo. P!).

Spathoglottis unguiculata auct. non (Labill.)Rchb.f. (1868).

S. angustifolia (Gaudich.)Benth. & Hook.f., Gen. Pl. 3: 512 (1883).

S. pacifica sensu Kraenzl. in Bot. Jahrb. Syst. 25: 603 (1898); H.Fleischm. & Rech. in Denkschr. Kaiserl. Akad. Wiss., Math.-Naturwiss. Kl. 85: 257 (1910); Christophersen in Bernice P. Bishop Mus. Bull. 128: 66 (1935), non Rchb.f.

S. spec. sensu Kraenzl. loc. cit.

S. daenikeri Kraenzl. in Viertelj. Naturf. Ges. Zürich 74: 80 (1929). Type: New Caledonia, *Daeniker* 1622 (holo. Z!).

A medium-sized to large terrestrial herb. Pseudobulbs small, ovoid, hidden by leaf bases. Leaves lanceolate to elliptic-lanceolate, acuminate, 40–90 × 2–6 cm, petiolate. Inflorescence erect, up to 1 m tall, densely many-flowered; bracts elliptic-lanceolate, acuminate, 1–2.5 cm long. Flowers showy, purple, pink or white with a yellow callus on the lip, pubescent on outer surface of sepals and petals; pedicel and ovary 3–5 cm long, densely shortly pubescent. Sepals elliptic to ovate, subacute to obtuse, 1.6–3 × 1–1.3 cm; laterals slightly oblique. Petals elliptic, obtuse, 2–3 × 1.3–1.8 cm. Lip 3-lobed at base, T-shaped when flattened; side lobes narrowly oblong, 0.7–0.8 cm long; midlobe spathulate, 1–1.3 cm long, auriculate at base, with a small callus on claw; callus between side lobes bilobulate, glabrous or sparsely hairy. Column clavate, incurved, 1.2–1.3 cm long.

Fig. 6. *Spathoglottis plicata*. **A**, habit × ⅕; **B**, flower × 1; **C**, dorsal sepal × 3; **D**, petal × 3; **E**, lateral sepal × 3; **F**, lip flattened × 2; **G**, lip in longitudinal section × 2; **H**, column ventral view × 2; **J**, anther cap two views × 4. **A** drawn from a colour slide; **B–H** drawn from *Hunt* 2939 by Sue Wickison.

DISTRIBUTION: 'Aunu'u, Ofu, Olosega, Savai'i, Ta'u, Tutuila, 'Upolu. Also widely distributed from Asia, S.E. Asia and Malay Archipelago to New Guinea, the Solomon Islands, Vanuatu, New Caledonia, the Horne Islands and Tonga.

HABITAT: Found on lava fields, waste places, by tracks and villages and in scrubby forest; sea level to 850 m

COLLECTIONS: *Bryan* 127 (BISH); *Christophersen* 230 (BISH, K), 597 (BISH), 677 (BISH, P), 1965 (BISH, P), 3552 (BISH); *Cox* 63 (BISH); *Garber* 710 (BISH), 712 (BISH), 1054 (BISH); *Graeffe* 80 (HAW); *Mitchell* 514 (BISH); *Rechinger* 1672 (W); *Reinecke* 185a (B†), 586 (B†); *Setchell* 193 (UC); *Vaupel* 476 (B, BISH, K); *Whistler* 1536 (BISH, HAW), 1956 (HAW), 2295 (HAW, K), 3459 (HAW), 6821 (HAW), 7576 (HAW), 8290 (HAW), 8558 (HAW); *Whitmee* s.n. (K, W); *Wilkes* s.n. (W); *Yuncker* 9007 (BISH), 9007a (BISH), 9416 (BISH), 9527 (BISH).

22. COELOGYNE

Lindl., Coll. Bot.: t.33 (1825).

Epiphytic or rarely lithophytic plants with conical, ovoid or angular pseudobulbs. Leaves terminal, 1–2, plicate or rarely appearing conduplicate, lanceolate, elliptic or ovate or oblanceolate. Inflorescences basal, spreading, erect or pendent, 1–many-flowered; bracts persistent or caducous. Flowers small to large, often showy. Sepals free, spreading, subsimilar. Petals free, often similar to sepals but usually smaller. Lip entire, 3-lobed or pandurate, usually with surface callosities, ecallose; side lobes erect when present; midlobe porrect to recurved; callus of warts, keels or ridges. Column elongate, clavate; pollinia 4, waxy.

A large genus of about 300 species in tropical and subtropical Asia, the Malay Archipelago, New Guinea, the Philippines and S.W. Pacific Islands. One species has been recorded from Samoa.

C. lycastoides *F.Muell. & Kraenzl.* in Oesterr. Bot. Zeit. 45: 179 (1895). Type: Samoa, *Betche* 24 (holo. MEL).
C. whitmeei Schltr. in Repert. Spec. Nov. Regni Veg. 11: 41 (1912). Type: Samoa, *Whitmee* s.n.(holo. B†, iso. K!).

A large epiphyte with narrowly ovoid pseudobulbs, 6–12 cm long, 1–2.5 cm in diameter, unifoliate at apex. Leaf lanceolate to elliptic, acute to acuminate, 25–60 × 12–16 cm, dark glossy green; petiole 2.5–5 cm long. Inflorescence spreading, up to 50 cm long, laxly 2–4-flowered, the flowers opening more or less simultaneously; bracts caducous, lanceolate, 3.8–6.5 cm long. Flowers pale green with rusty brown markings on the lip; pedicel and ovary 6-winged, 2–3 cm long. Sepals elliptic-ovate to oblong-ovate, acute, 3.5–4.8 × 1.1–1.6 cm; laterals somewhat oblique and keeled on outer surface. Petals linear, acute,

3.5–4.8 × 0.3–0.4 cm. Lip 3-lobed, 3.2–4 × 2–2.6 cm; side lobes small, erect, semi-elliptic or semi-circular; midlobe oblong-obovate, recurved, obtuse to abruptly acuminate; callus of several verrucose lines, the outer two extending onto middle of midlobe. Column 2–2.4 cm long.

DISTRIBUTION: Savai'i, Ta'u, Tutuila, 'Upolu. Also in Vanuatu and New Caledonia.

HABITAT: Epiphyte in rain-forest and cloud-forest; 300–1550 m.

COLLECTIONS: *Betche* 24 (MEL); *Christophersen* 175 (BISH), 179 (BISH), 201 (BISH), 367 (BISH), 720 (BISH), 1021 (BISH), 1068 (BISH), 1964 (BISH), 2269 (BISH), 2298 (BISH, P); *Cox* 211 (BISH), 346 (BISH); *Garber* 788 (BISH); *Powell* 254 (K); *Vaupel* 460 (B†); *Whistler* 156 (BISH, HAW), 410 (HAW), 521(BISH, HAW), 1400 (HAW), 2612 (BISH, HAW), 2796 (BISH, HAW), 7006 (HAW), 8648 (HAW), s.n. (K); *Whitmee* s.n.(K).

23. **LIPARIS**

L.C.Rich., Orch. Europ. Annot.: 21, 30, 38 (1817) & in Mem. Mus. Hist. Nat. Paris 4: 43, 52, 60 (1818).

Small to medium-sized, terrestrial, lithophytic or epiphytic herbs. Stems pseudobulbous, short to long, covered when young by sterile bracts, leafy. Leaves one to several, linear to ovate or elliptic, pleated or not, thin-textured to somewhat coriaceous. Inflorescence terminal, erect, racemose, few to many-flowered. Flowers usually small, yellow, green, orange or purple, often somewhat translucent, resupinate. Sepals free, spreading; laterals sometimes fused for part or all of length. Petals free, often linear and unlike sepals, often reflexed. Lip often strongly reflexed, ovate, oblong or flabellate, entire or lobed, usually with a basal callus, lacking a spur. Column incurved, clavate, elongate; pollinia four in two pairs.

A large cosmopolitan genus of some 350 species, well represented in tropical Asia, the Malay Archipelago, the Philippines, New Guinea, Australia and the S.W. Pacific Islands, with five species in Samoa.

1. Leaf solitary · 1
   Leaves two or more · 4
2. Leaf heart-shaped · **L. phyllocardia**
   Leaf linear · 3
3. Rachis very short in comparison with peduncle; bracts imbricate and distichous; flowers orange · **L. gibbosa**
   Rachis similar to peduncle in length; flowers pale yellow or greenish yellow; bracts not distichous and imbricate · · · · · · · · · · · **L. caespitosa**
4. Leaves linear or oblanceolate, flowers greenish · · · · · **L. condylobulbon**
   Leaves ovate; flowers purple with a green column · · · · · · · · · **L. layardii**

L. caespitosa *(Thouars) Lindl.* in Bot. Reg. 2, sub t. 882 (1825). Type: Mauritius, *Thouars* s.n. (holo. P!).

Malaxis caespitosa Thouars in Orch. Iles. Austr. Afr.: t.90 (1805).

Leptorchis caespitosa (Thouars) Kuntze, Rev. Gen. Pl.: 671 (1891).

Malaxis angustifolia Blume, Bijdr. Fl. Ned. Ind.: 393 (1825). Type: Java, *Blume* s.n. (holo L!).

Liparis angustifolia (Blume) Lindl., Gen. Spec. Orch. Pl.: 31 (1830).

L. neoguineensis Schltr. in Repert. Spec. Nov. Regni Veg. Beih. 1: 209, fig.289 (1911). Type: New Guinea, *Schlechter* 13934 (holo. B†, iso. K!).

L. sp. p.p. sensu Christophersen in Bernice P. Bishop Mus. Bull. 128: 64 (1935).

A small epiphyte with clustered ovoid, slightly compressed pseudobulbs, 0.8–1.5 cm long, 0.6–1.3 cm in diameter, unifoliate. Leaf erect, oblanceolate to linear-lanceolate, acute or subacute, 6–17 × 1–1.4 cm; petiole short. Inflorescence erect, 6–17 cm long, laxly to subdensely many-flowered; peduncle slightly flattened, slender; rachis about as long as the peduncle; bracts linear-lanceolate, acuminate, 5 mm long. Flowers pale yellow-green; pedicel and ovary 3–4.5 mm long, weakly 3-angled. Sepals elliptic-ovate, acute, 2–2.6 × 0.8–1.1 mm. Petals linear, acute or subacute, 2–2.4 × 0.3–0.4 mm. Lip recurved, elliptic or oblong-ovate, obtuse, 2–2.5 × 1.7–2 mm, lacking a callus. Column 2–2.2 mm long, slightly incurved.

DISTRIBUTION: Tutuila. Also in Tropical Africa, Madagascar, the Mascarene Islands, Sri Lanka, India, S.E.Asia, the Malay Archipelago, New Guinea, Bougainville, the Solomon Islands, Vanuatu, Fiji and across to the Austral Islands.

HABITAT: Rare epiphyte found on tree trunks in montane forest; c.500 m.

COLLECTIONS: *Christophersen* 1182 (BISH), 1200 (BISH).

L. condylobulbon *Rchb.f.* in Hamburg. Garten-Blumenzeit. 18: 34 (1862). Type: Burma, Moulmein, *Parish* (holo. W!).

L. longipes Kraenzl. in Bot. Jahrb. Syst. 25: 600 (1898), non Lindl.

L. nesophila Rchb.f. in Otia Bot. Hamburg. 1: 56 (1878). Type: Fiji, *Seemann* 615 (holo. K!, iso. AMES!, P!).

L. confusa J.J.Sm., Orch. Java: 275 (1905). Type: Java, *J.J.Smith* 903 (holo. BO!).

L. savaiiensis H.Fleishm. & Rech. in Denkschr. Kaiserl. Akad. Wiss., Math.-Naturwiss. Kl. 85: 255, t.1, fig.4 (1910). Type: Samoa, Savai'i, *Rechinger* 1889 (holo. W!).

L. elegans sensu Ames in J. Arnold Arb. 14: 105 (1933).

Epiphytic herb, 15–40 cm tall. Pseudobulbs slender, 7–20 × 0.8–1.4 cm, two-leaved towards the apex. Leaves suberect, linear-oblanceolate, acute, 10–25 × 1–2.8 cm. Inflorescence 6–20 cm long, densely many-flowered; rachis 5–15 cm long; bracts lanceolate, acuminate, 3–4 mm long. Flowers small, probably self-pollinating, pale green, often with a brown or reddish lip; pedicel and ovary 4–8 mm long. Dorsal sepal elliptic, subacute, 2.5–3.5 × 1–1.7 mm. Lateral

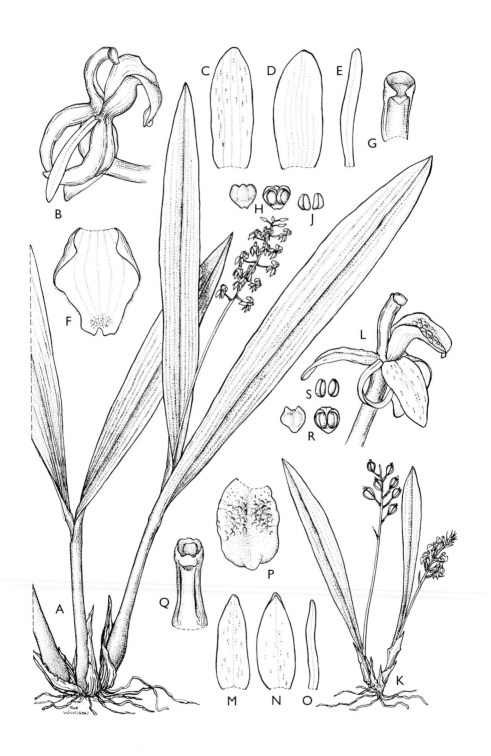

sepals elliptic, obtuse, 2.5–3.5 × 1–2 mm. Petals linear, 2–3 × 0.25–0.5 mm. Lip oblong or elliptic-oblong, retuse, strongly reflexed, 2.5–3 × 1.5–2 mm; callus bilobed, at base of lip. Column slightly incurved at apex, 1.5–2 mm long.

DISTRIBUTION: Ofu, Savai'i, Ta'u, Tutuila, 'Upolu. Also occurring from S.E. Asia into Malesia, and eastwards to the Solomon Islands, Vanuatu, New Caledonia and Fiji.

HABITAT: Epiphyte found in coastal to montane forests; sea level to 700 m.

COLLECTIONS: *Christophersen* 109 (BISH, K), 1263 (BISH), 1835 (BISH, K), 1949 (BISH, P); *Cox* 306 (BISH), 339 (BISH); *Garber* 848 (BISH, P); *Powell* 358 (K), 359 (K), 360 (K); *Rechinger* 123 (W), 207 (W), 1506 (W), 1889 (W); *Vaupel* 157 (K); *Whistler* 31 (HAW), 210 (BISH, HAW), 545 (HAW), 563 (BISH, HAW), 1185 (BISH, HAW), 1559 (HAW), 1976 (BISH, HAW), 2699 (HAW), 2739 (BISH, HAW), 3005 (BISH), 7057 (HAW), 7893 (HAW), 8150 (HAW), 9040 (HAW); *Whitmee* s.n. (K); *Wild* 15 (BISH); *Wisner* 138 (BISH).

L. gibbosa *Blume ex Finet* in Bull. Soc. Bot. France 55: 342, t.11.f.36–44 (1908). Type: Java, *Blume* s.n. (holo. P!).
Malaxis gibbosa Blume nom. in sched.
Liparis disticha sensu Schltr. in Bot. Jahrb. Syst. 39: 60 (1906); non Lindl.
L. sp. p.p. sensu Christophersen in Bernice P. Bishop Mus. Bull. 128: 64 (1935).

An epiphytic or rarely lithophytic herb. Pseudobulbs well spaced on rhizome, ovoid, 1–2 cm long, 1–1.5 cm in diameter, unifoliate. Leaf erect, linear, acute, 15–23 × 0.8–1.2 cm, shortly petiolate. Inflorescence erect, 6–18 cm long; peduncle slender, somewhat flattened, winged above; rachis short, flattened, 1–4 cm long; bracts distichous, imbricate, conduplicate, ovate, acuminate. Flowers successive, yellow-green to pale orange with a darker orange-brown lip; pedicel and ovary 1.2–1.5 cm long, weakly 6-ribbed. Sepals reflexed, oblong-elliptic to ovate, abruptly acuminate, 5–6 × 2.2–2.4 mm. Petals erect, oblanceolate, subacute, 4.5–6 × 1–1.5 mm. Lip strongly recurved, more or less orbicular-ovate, acute, 4–4.5 × 3.5–4 mm; callus basal, obscure, bilobulate. Column dilated at base, winged at apex, 2–2.5 mm long.

DISTRIBUTION: Savai'i, Tutuila. Also widely distributed from S.E. Asia, the Malay Archipelago to New Guinea, the Solomon Islands, Vanuatu, New Caledonia and Fiji.

HABITAT: Epiphyte found in montane scrub; 500–620 m.

COLLECTIONS: *Christophersen* 1062 (BISH), 1187 (BISH), 1191 (BISH); *Whistler* 9635 (HAW, K).

Fig. 7. *Liparis condylobulbon*. **A**, habit × ⅔; **B**, flower in side view × 10; **C**, dorsal sepal × 10; **D**, lateral sepal × 10; **E**, petal × 10; **F**, lip × 10; **G**, column × 14, **H**, anther (two views) × 14; **J**, pollinia × 4. *L. caespitosa*. **K**, habit × ⅔; **L**, flower × 10; **M**, dorsal sepal × 10; **N**, lateral sepal × 10; **O**, petal × 10; **P**, lip × 10; **Q**, column from beneath × 14; **R**, anther cap (two views) × 14; **S**, pollinia × 14. **A** drawn from *Hallé* 6394; **B–J** from *Wickison* 50; **K–S** from *Cribb et al.* 5043. All drawn by Sue Wickison.

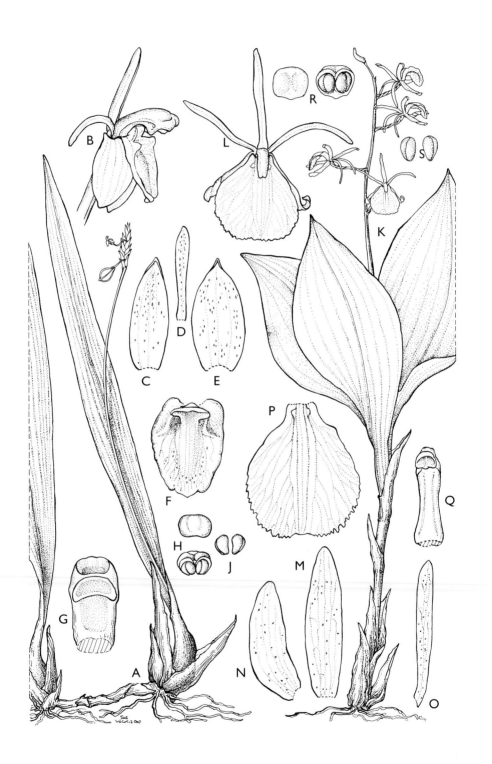

L. layardii *F.Muell.* in South Sci. Rec.: 1 (1885). Type: New Caledonia, *Layard* s.n. (holo. MEL!).

L. stricta Schltr. in Repert. Spec. Nov. Regni Veg. 9: 95 (1910). Type: Samoa, *Vaupel* 134 (holo. B†, iso. K!).

L. mataanensis J.J.Sm. in Bull. Jard. Bot. Buitenzorg 8: 56 (1912). Type: As for *L. stricta.*

L. sp. 1 sensu Kraenzl. in Bot. Jahrb. Syst. 25: 601 (1898).

Erect terrestrial plants up to 45 cm tall. Pseudobulbs clustered, cylindrical, up to 20 cm long, 1 cm in diameter, 2–3-leaved. Leaves ovate to ovate-elliptic, acute, 6.5–10 × 5–7 cm; petiole 2–3.5 cm long. Inflorescence laxly 8–15-flowered, 12–30 cm long; bracts 5–10 mm long. Flowers purple with a green column; pedicel and ovary 1–1.2 cm long, 6-angled. Dorsal sepal erect, oblong-lanceolate, acute, 7–12 × 2–2.5 mm; lateral sepals reflexed, oblong to oblong-ovate, obtuse, 6.5–10 × 2.5–3 mm. Petals linear-ligulate, subacute, 6–12 × 0.75–1.25 mm. Lip recurved, flabellate to obovate, 6.5–10 × 5.5–7.5 mm, the margins weakly crenulate; callus bilobulate, basal. Column incurved at apex, 5–6 mm long.

DISTRIBUTION: Ofu, Olosega, Savai'i, Ta'u, 'Upolu. Also in the Solomon Islands, Vanuatu, New Caledonia and Fiji.

HABITAT: Terrestrial orchid found in rain-forest; 340–1700 m.

COLLECTIONS: *Christophersen* 778 (BISH), 821 (BISH), 891 (BISH), 2207 (BISH), 3445 (BISH); *Reinecke* 290 (B†), 290a (B†); *Vaupel* 134 (K); *Whistler* 2561 (HAW), 2635 (BISH, HAW), 3098 (HAW), 3192 (HAW), 3695 (HAW), 5716 (HAW), 6891 (HAW), 7830 (HAW, 9601 (HAW), 9550 (HAW).

NOTE: Possibly the same as *L. longa* Rchb.f., an earlier name.

L. phyllocardia *Schltr.* in Repert. Spec. Nov. Regni Veg. 9: 94 (1910). Type: Samoa, *Vaupel* 393 (holo. B †).

L. sp. 2 sensu Kraenzl. in Bot. Jahrb. Syst. 25: 601 (1898).

L. phyllocardia Schltr. var. *minor* Schltr. in Repert. Spec. Nov. Regni Veg. 9: 95 (1911). Type: Samoa, Savai'i, *Vaupel* 11 (holo. B†).

A terrestrial herb 15–25 cm tall. Pseudobulb approximate, ovoid-ellipsoidal, 2–3 × 1–1.5 cm, unifoliate. Leaf heart-shaped, 8.5–18.5 × 6–13.5 cm, longly petiolate; petiole 5–9 cm long. Inflorescence erect, densely 10–20-flowered, racemose to subcapitate; bracts linear-lanceolate, acuminate, 3–10 mm long. Flowers whitish, possibly cleistogamous; pedicel and ovary 17–23 mm long.

Fig. 8. *Liparis gibbosa*. **A**, habit × 1; **B**, flower × 6; **C**, dorsal sepal × 8; **D**, petal × 8; **E**, lateral sepal × 8; **F**, lip × 8; **G**, column × 10; **H**, anther cap × 10; **J**, pollinia × 10. *L. layardii*. **K**, habit × 0.5; **L**, flower × 3; **M**, dorsal sepal × 3; **N**, lateral sepal × 3; **O**, petal × 3; **P**, lip × 3; **Q**, column from below × 3; **R**, anther (two views) × 8; **S**, pollinia × 8. A–J drawn from *Wickison* 105 A; **K** from *Mackee* 19; **L–S** from *Mitchell* 5. All drawn by Sue Wickison.

Dorsal sepal lanceolate, subacute, 10 mm long. Lateral sepals triangular-lanceolate, obtuse, 10 mm long. Petals linear, subacute, 10 mm long. Lip broadly obovate, apiculate, 10 × 8 mm, with a serrulate margin; basal callus very small. Column a little dilated at the base, semiterete above, 4 mm long.

DISTRIBUTION: Savai'i. Endemic.

HABITAT: Terrestrial orchid found in montane and cloud-forest; 1100–1700 m.

COLLECTIONS: *Christophersen* 777 (BISH), 822 (BISH), 2126 (BISH), 2191 (BISH); *Powell* 252 (K, W); *Reinecke* s.n. (B†); *Vaupel* 393 (B†); *Whistler* 2506 (BISH, HAW), 2522 (HAW); *Whitmee* s.n. (K).

24. **MALAXIS**

Solander ex Sw., Nov. Gen. Sp. Prodr.: 8 (1788).

Terrestrial or rarely epiphytic herbs, with hairy roots. Stems cylindrical to pseudobulbous, fleshy, often creeping and rooting in basal part, leafy. Leaves 1 to several, thin-textured to fleshy, pleated, petiolate; petiole sheathing at base. Inflorescence apical, erect, racemose, laxly to densely few to many-flowered. Flowers small, green, brown, yellow, pink or purple, non-resupinate. Dorsal sepal spreading free. Lateral sepals free or fused, spreading. Petals often narrower than sepals, free, spreading. Lip erect, flat but sometimes concave at base, entire to lobed, auriculate at base, apical margins often toothed, lacking a spur, callus absent or obscurely ridged. Column very short, lacking a foot; pollinia 4, waxy, lacking appendages.

A genus of about 300 species found throughout the tropics and subtropics with a few species in temperate regions. Five species have been reported in Samoa. The taxonomy of the Pacific Island species is in urgent need of revision.

1. Lip 4-lobed or 3-lobed with midlobe deeply retuse; flowers green
 · **M. tetraloba**
 Lip 3-lobed or with lateral teeth on either side of small midlobe · · · · · **2**
2. Leaves lanceolate, less than 1.6 cm wide; flowers off-white
 · **M. samoensis**
 Leaves ovate, widest more than 1.8 cm wide · · · · · · · · · · · · · · · · · **3**
3. Leaves basal, 3–5 · **M. resupinata**
 Leaves 7 or more arranged along the elongate erect stem · · · · · · · · · **4**
4. Leaves broadly ovate, less than twice as long as broad; flowers pale yellow; lip broader than long, the side lobes with several teeth along front margin · **M. reineckeana**
 Leaves narrowly ovate, twice as long as broad; flowers purple or pale yellow; lip longer than broad; side lobes acute but not toothed in front
 · **M. taurina**

M. reineckeana *(Kraenzl.) Kores* in Allertonia 5(1): 51 (1989). Types: Samoa, *Reinecke* 311 & 620 (syn. B†) .

Microstylis reineckeana Kraenzl. in Bot. Jahrb. Syst. 25: 600 (1898).

Microstylis whitmeei Rolfe in Bull. Misc. Inform., Kew 1922: 23 (1922). Type: Samoa, *Whitmee* s.n. (holo. K!).

Malaxis whitmeei (Rolfe) Kores in Allertonia 5(1): 52 (1989).

A medium-sized to large terrestrial herb with an elongate erect stem up to 20 cm long, leafy along its length, growing from an elongate fleshy stem. Leaves up to 12, distichous and twisted to lie in one plane, ovate, acute, 7–11 × 3–5 cm; petiole sheathing at base, up to 4 cm long. Inflorescence densely many-flowered, up to 30 cm long; bracts reflexed, linear, acuminate, 5–6 mm long. Flowers pale yellow, may be cleistogamous; pedicel and ovary 5 mm long. Dorsal sepal oblong, obtuse, 3.5–4 × 1.5–1.8 mm. Lateral sepals similar to dorsal sepal. Petals oblong, obtuse, 2.5–3 × 1 mm. Lip transversely reniform-subcircular, obscurely 3-lobed, 4–4.5 × 5–6 mm; midlobe small, emarginate; side lobes rounded, shortly 3-toothed in front. Column short.

DISTRIBUTION: Savai'i, Tutuila, 'Upolu. Also in the Solomon Islands and the Cook Islands.

HABITAT: Terrestrial orchid found in montane forest. 300–750 m.

COLLECTIONS: *Bryan* 91 (BISH); *Christophersen* 564 (BISH), 1128 (BISH), 3582 (BISH); *Faasuaga* 9209 (HAW); *Long* 3026 (HAW); *Rechinger* 436 (W), 651 (W); *Reinecke* 311 (B†), 620 (B†); *Vaupel* 279 (K); *Whistler* 567 (HAW), 729 (HAW), 1580 (HAW), 1639 (HAW), 1651 (HAW), 3236 (HAW, K), 4688 (HAW), 7007 (HAW), 7015 (HAW), 7046 (HAW), 7070 (HAW), 7093 (HAW), 9490 (HAW), 9550 (HAW); *Whitmee* s.n. (K); *Wilkes* s.n. (W); *Wisner* 141 (BISH).

M. resupinata *(G.Forst.) Kuntze*, Rev. Gen. Pl. 2: 673 (1891). Type: Tahiti, *J.R. & G.Forster* s.n. (holo. BM!).

Epidendrum resupinatum G.Forst., Fl. Ins. Austr. Prodr.: 61 (1786).

Microstylis resupinata (G.Forst.) Drake, Ill. Fl. Ins. Mar. Pac.: 305 (1892).

Microstylis reineckeana sensu Christophersen in Bernice P. Bishop Mus. Bull. 128: 63 (1935) non Kraenzl.

Terrestrial plants with decumbent or ascending terete stems, 5–10 cm long, 3–5-leaved. Leaves erect or ascending, obliquely oblong-elliptic or ovate-elliptic, acuminate, 10–22 × 4.5–8 cm; petiole relatively stout, 6–10.5 cm long, somewhat dilated and sheathing at base. Inflorescence erect, 21–45 cm long, laxly many-flowered; peduncle weakly angular, 15–25 cm long; bracts lanceolate, acuminate, 2.5–5 mm long. Flowers maroon; pedicel and ovary c. 6 mm long. Dorsal sepal oblong to oblong-elliptic, subacute, 3.5–4.5 × 1–1.5 mm. Lateral sepals falcate, oblong, obtuse, 3–4 mm long, 1.5 mm. Petals slightly falcate, linear-ligulate, subacute, 3.5–4 × 0.7–0.8 mm wide. Lip more or less oblong-obovate, prominently auriculate at base, weakly 3-lobed in front, 5–6 × 4–5 mm; side lobes broadly rounded with several short teeth on front margin; midlobe small, semiorbicular, acutely bilobed; callus horseshoe-shaped. Column 0.5–1 mm long.

DISTRIBUTION: Ofu, Savai'i, Ta'u, Tutuila, 'Upolu. Also widely distributed in the South Pacific, Vanuatu, Fiji, Tonga, the Cook Islands, and the Society Islands.

HABITAT: Terrestrial orchid found in rain-forest; 250–500 m.

COLLECTIONS: *Christophersen* 3230 (BISH); *Setchell* 387 (UC); *Vaupel* 597 (B†); *Whistler* 2727 (HAW), 3018 (HAW), 3099 (HAW), 3475 (HAW), 5709 (HAW), 7597 (HAW), 7928 (HAW), 7930 (HAW), 8778 (HAW); *Yuncker* 9431 (BISH), 9522 (BISH).

M. samoensis *(Schltr.)Whistler* in Phytologia 38(5): 409 (1978). Type: Samoa, 'Upolu, *Betche* s.n. (holo. B†).

Microstylis samoensis Schltr. in Repert. Spec. Nov. Regni Veg. 9: 93 (1910).

Microstylis. vitiensis Schltr. in Repert. Spec. Nov. Regni Veg. 10: 249 (1911). Type: Fiji, *Lucae* s.n. (holo. B†).

Microstylis schlechteri Rolfe in Bull. Misc. Inform., Kew 1921: 53 (1921). Nom. nov. pro *Microstylis vitiensis* Schltr.

Malaxis schlechteri (Rolfe) L.O.Williams in Bot. Mus. Leafl. 5: 115 (1938).

Liparis sp. sensu Yuncker in Bernice P. Bishop Mus. Bull. 184: 31 (1945).

A small terrestrial herb with short cylindrical stems up to 3.5 cm long. Leaves 3–4, suberect, lanceolate or narrowly lanceolate, acuminate, 5–11.5 × 0.8–1.6 cm, slenderly petiolate at base. Inflorescence erect, laxly few-flowered, 12–15 cm long; peduncle; bracts lanceolate, acuminate, 4–5 mm long, reflexed. Flowers white or pale greenish; pedicel and ovary 4–5 mm but elongating after fertilisation. Dorsal sepal oblong, obtuse, 3 × 1.5 mm. Lateral sepals obliquely oblong, obtuse, 3 mm long, 1.3 mm wide. Petals linear, obtuse, 2.5–3 × 0.5 mm. Lip erect, broadly subcircular, obscurely 3-lobed, 4 mm long and wide; side lobes rounded, 3–4-toothed in front; midlobe oblong, deeply emarginate, each lobule acuminate; callus obscurely horseshoe-shaped. Column short.

DISTRIBUTION: Olosega, Ta'u, Tutuila, 'Upolu. Also in Fiji.

HABITAT: Terrestrial orchid found in rain- and cloud-forests; 300–800 m.

COLLECTIONS: *Betche* s.n. (B†); *Christophersen* 3567 (BISH); *Whistler* 2742 (BISH, HAW, K), 3082 (HAW), 3193 (HAW), 3250 (HAW), 3730 (HAW), 4701 (HAW), 5725 (HAW), 7731 (HAW), 7929 (HAW), 8507 (HAW), 8716 (HAW), 8994 (HAW), s.n. (K); *Yuncker* 9266 (BISH).

M. taurina *(Rchb.f.)O.Kuntze*, Rev. Gen. Pl.: 573 (1891). Type: New Caledonia, *Deplanche* s.n. (holo. W!).

Microstylis taurina Rchb.f. in Linnaea 41: 97 (1877).

A terrestrial herb with short erect stems 3–6(–15 cm) long. Leaves 5–10, ovate, acute, 6–9 × 1.8–3.5 cm; petiole slender, up to 5 cm long, sheathing at the base. Inflorescence subdensely many-flowered; peduncle 11–15 cm long; bracts reflexed, linear, acuminate, 4–6 mm long. Flowers small, purple or rarely dull yellowish; pedicel and ovary 3–4 mm long. Dorsal sepal ovate, obtuse, 3 mm long, 2 mm wide. Lateral sepals oblong, obtuse, 3 × 1.5 mm. Petals linear-oblong, obtuse, 3 × 1 mm. Lip 3-lobed, 4 × 3.5–4 mm; side lobes hatchet-shaped, acute in front; midlobe oblong-elliptic or tapering, emarginate. Column short.

DISTRIBUTION: Ta'u. Vanuatu, New Caledonia and Fiji.
HABITAT: Terrestrial orchid in montane forest; 300–800 m.
COLLECTIONS: *Garber* 753 (BISH); *Whistler* 1405 (HAW), 3193B (HAW), 3711 (HAW).

M. tetraloba *(Schltr.) Kores* in Allertonia 5(1): 50 (1989). Type: Samoa, *Betche* s.n. (holo. B†).
Microstylis tetraloba Schltr. in Repert. Spec. Nov. Regni Veg. 9: 94 (1910).
M. radicicola Rolfe in Bull. Misc. Inform., Kew 1921: 53 (1921). Type:
Malaxis radicicola (Rolfe) L.O.Williams in Bot. Mus. Leafl. 5: 115 (1938).

A terrestrial or epiphytic plant 10–30 cm tall. Stems clustered on a short fleshy rhizome, conical at base, 4–7 × 1–1.3 cm, densely 3–5-leaved. Leaves erect or ascending, elliptic to elliptic-ovate, acuminate, 8–16 × 3.3–6 cm, with a slender petiole, 0.8–1.2 cm long, dilated and sheathing at base. Inflorescence erect, 12–25 cm long, densely many-flowered; peduncle terete, stout, 7–17 cm long; bracts weakly reflexed, linear-tapering, 3–5 mm long. Flowers green to pale green, pedicel and ovary 5–6 mm long. Dorsal sepal oblong-elliptic, subacute, 4–5 × 1.5–1.75 mm. Lateral sepals broadly elliptic to elliptic-obovate, obtuse, 3.5–4.5 × 2–2.5 mm. Petals spreading or weakly reflexed, linear-ligulate, subacute, 3.5–4 × 0.7–0.8 mm. Lip oblong-pandurate, obscurely 4-lobed, very prominently auriculate at base, 7–7.5 × 6 mm, the apical lobules oblong-obovate, the anterior margins weakly crenulate, broadly rounded in front; callus weakly horseshoe-shaped. Column c.1 mm long.

DISTRIBUTION: Savai'i and 'Upolu. Also from Fiji.
HABITAT: A terrestrial orchid found in rain-forest and cloud-forest; 380–700 m.
COLLECTIONS: *Betche* s.n. (B†); *Sledge* 1584 (K); *Whistler* 564 (HAW), 785 (HAW), 1511 (HAW), 4700 (HAW), 6939 (HAW).

25. **OBERONIA**
Lindl., Gen. Spec. Orch. Pl.: 15 (1830)

Small to large epiphytic plants with short to long leafy stems, lacking pseudobulbs. Leaves iridiform, equitant, short to long, often fleshy, articulated or not at base, distichous, often imbricate at base. Inflorescence terminal, laxly to densely many-flowered, pubescent or glabrous. Flowers often in whorls, small, non-resupinate, flat. Sepals and petals free, spreading. Lip larger, sessile, entire or lobed, usually spreading, occasionally somewhat concave at base. Column short; anther terminal; pollinia 4, waxy, cohering in two pairs.

A genus of some 150 to 200 species centred on tropical South and S.E. Asia and the Malay Archipelago but extending to tropical Africa, Madagascar, the Mascarene Islands, the Philippines, New Guinea, Australia and the S.W. Pacific Islands across to Tahiti. Three species have been reported from Samoa.

1. Stem very short; leaves much longer than stem, up to 40 cm long
    · **O. heliophila**
    Stem elongate; leaves shorter than the stem, 2–6 cm long · · · · · · · · · **2**
2. Leaves 1–1.5 cm wide · **O. equitans**
    Leaves 0.2–0.3 cm wide · **O. bifida**

O. bifida *Schltr.* in Schum. & Lauterb., Nachtr. Fl. Schutzgeb. Sudsee: 111(1905). Type: New Guinea, *Schlechter* 14085 (holo. B†, iso. K!).

An epiphyte with clustered elongate stems up to 10 cm long, sinuous in longest ones. Leaves equitant, linear-lanceolate, acuminate, 5–8 × 0.2–0.3 cm. Inflorescence cylindrical, densely many-flowered, 3–10 cm long. Flowers verticillate, pale pinkish brown; ovary 1.5 mm long. Sepals ovate, obtuse, 0.7–0.8 × 0.5–0.6 mm. Petals linear-ovate, subacute, 0.7 × 0.3 mm. Lip oblong, deeply retuse in front, 1 mm × 0.5 mm, apical lobes obtuse. Column very short.

DISTRIBUTION: Savai'i. Also in New Guinea, the Solomon Islands, Vanuatu and New Caledonia.

HABITAT: Epiphyte in montane forest; 700–800 m.

COLLECTIONS: *Christophersen* 2296b (BISH).

NOTE: This is a tentative identification. The specimen cited here is fruiting but it matches well the type material and specimens from the Solomon Islands in its habit. The description of the flowers is taken from Solomon Islands material.

O. equitans *(G.Forst.)Mutel* in Premier Mém. sur les Orch. Paris: 8 (1838).Type: Tahiti, *G.Forster* 170 (holo. BM!, iso. P!).

Epidendrum equitans G.Forst., Fl. Ins. Austr. Prodr.: 60 (1786).

Cymbidium equitans (Forst. f.)Sw. in Nova Acta Regiae Soc. Sci. Upsal. 6: 72 (1799).

Oberonia glandulosa Lindl., Fol. Orch., Oberonia: 6 (1859). Type: Pacific Islands, *Matthews* 158 (holo. K, not traced)

Malaxis glandulosa (Lindl.) Rchb.f. in Ann. Bot. Syst. 6: 215 (1861).

Oberonia brevifolia sensu Seem. in Bonplandia 10: 153 (1861), non Lindl.

O. flexuosa Schltr. in Bot. Jahrb. Syst. 39: 61 (1906). Type: New Caledonia, *Schlechter* 15496 (holo. B†).

O. palmicola sensu B.E.V.Parham in Trans. Proc. Fiji Soc. 2: 27 (1953), non F.Muell.

Malaxis equitans sensu N. Hallé in Fl. Nouv-Caled. Dépend. 8: 270, prosyn. (1977), non Blume.

Oberonia aff. *diura* sensu Christophersen in Bernice P. Bishop Mus. Bull. 128: 63 (1935).

Small epiphytic plants, 14–21 cm long. Stem elongate, flexuose, covered in imbricate leaf-bases, up to 14 cm long. Leaves distichous, imbricate, equitant, narrowly lanceolate-falcate, acute, 2–6 × 0.4–0.8 cm. Inflorescence cylindrical, up to 10 cm long, densely many-flowered, minutely pubescent; bracts subdeltoid, 1.5–2 mm long, erosulate. Flowers verticillate, greenish cream to cream-coloured; ovary 1.5 mm long, papillate. Dorsal sepal oblong, obtuse, c.1 mm long, sparsely papillate on outer surface. Lateral sepals similar. Petals oblong-ovate, acute to subacute, 0.8 mm long. Lip subentire, subquadrate, obscurely bilobulate at apex, biauriculate at base, 1 mm long. Column very short.

DISTRIBUTION: Ofu, Olosega, Savai'i, Tutuila, 'Upolu. Also in the Solomon Islands, Vanuatu, New Caledonia, Fiji, Tonga, Tahiti and Tuamotus.

HABITAT: Epiphyte found in lowland and montane forest; sea level–1550 m.

COLLECTIONS: *Christophersen* 172 (BISH, K), 1039 (BISH); *Dickie* 9 (K); *Garber* 880 (BISH); *Graeffe* s.n. (W); *Long* 3087 (HAW); *Rechinger* 16 (W), 127 (W), 679 (W), 1281 (W), 1589 (W); *Reinecke* 184 (B†); *Setchell* 210 (BISH, UC), 336 (UC); *Vaupel* 159 (B†); *Whistler* 204 (BISH, HAW), 532 (HAW), 2638 (BISH, K, HAW), 2906 (HAW, K), 3012 (BISH, HAW), 3809 (BISH, HAW), 3946 (BISH, HAW), 8666 (HAW), 9039 (HAW); *Wilkes* s.n. (W).

O. heliophila *Rchb.f.* in Otia Bot. Hamburg.: 56 (1878). Type: Samoa, Savai'i, *U.S. Expl. Exped.* (lecto. W!, isolecto. AMES!, US).

Malaxis heliophila Rchb.f. in Otia Bot. Hamburg.: 56, nom. alt. (1878). Type: Samoa, *Whitmee* s.n. (holo. K!, iso. BM!).

Oberonia betchei Schltr. in Bull. Herb. Boissier 2, 6: 303 (1906). Type: Samoa, *Betche* 38 (holo. B†, iso. MEL!).

O. iridifolia sensu Kraenzl. in Bot. Jahrb. Syst. 25: 601 (1898); H.Fleischm. & Rech. in Denkschr. Kaiserl. Akad. Wiss., Math.-Naturwiss. Kl. 85: 256 (1910), non Lindl.

O. verticillata sensu Kraenzl. loc. cit.; sensu H.Fleischm. & Rech. loc. cit., non Wright

A very short stemmed epiphyte up to 55 cm long. Leaves distichous, imbricate at base, equitant, linear to lanceolate, slightly falcate, acuminate, 14–40 × 0.7–1.4 cm. Inflorescence 25–45 cm long, densely to sparsely tomentose, densely many-flowered; bracts oblong-ovate, erose on margins. Flowers subverticillate, yellow to orange-yellow, the lip darker; ovary 1 mm long, glabrous. Dorsal sepal oblong-ovate, subacute, 1 × 0.7 mm. Lateral sepals obliquely oblong-ovate, obtuse, 1 mm long. Petals ligulate-lanceolate, obtuse, 1 mm long. Lip entire, oblong-subquadrate, retuse or truncate at apex, c. 1.2 mm long, shortly biauriculate at the base. Column very short.

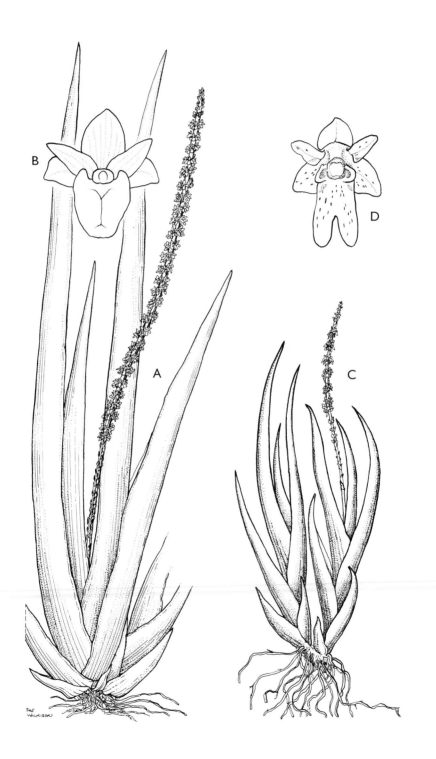

DISTRIBUTION: Apolima, Savai'i, Tutuila, 'Upolu. Also the Solomon Islands, Vanuatu and Fiji.

HABITAT: Epiphyte found from the coastal to montane forests; near sea-level to 550 m.

COLLECTIONS: *Betche* 38 (MEL); *Christophersen* 146 (BISH, P), 574 (BISH, K), 1882 (BISH, K); *Diefenderfer* 22 (BISH); *Graeffe* 1257 (HAW), s.n. (W); *Long* 3100 (BISH, HAW); *Rechinger* 99 (W), 1202 (W); *Reinecke* 214 (B†), 240 (B†), 615 (B†), 628 (B†); *Vaupel* 234 (BISH, HAW, K); *Whistler* 209 (HAW), 1702 (HAW, K), 2790 (BISH, HAW, K), 3947 (BISH, HAW), 5350 (HAW), 8153 (HAW), 9039 (HAW); *Whitmee* 45 (K), 165 (K), s.n. (BM, K); *Wilkes* s.n. (AMES, US, W); *Yuncker* 9307 (BISH).

26. **ERIA**

Lindl. in Bot. Reg.11: t.904 (1825).

Small to large epiphytic or rarely terrestrial plants with pseudobulbous stems. Pseudobulbs 1–several-noded, cylindrical to ovoid or conical, usually 2- or more-leaved in apical part, covered by thin cataphylls when young. Leaves fleshy to coriaceous, linear to lanceolate, oblong or ovate. Inflorescences elongate, laxly to densely spicate or racemose, two–many-flowered, from upper nodes or terminal. Flowers resupinate, often covered in hairs. Dorsal sepal and petals free. Lateral sepals oblique, connate at base with column-foot to form a more or less distinct mentum. Lip attached to the apex of the column foot, entire to 3-lobed, with or without a callus on the disc, lacking a spur. Column short, stout, with a more or less prominent foot; anther terminal; pollinia 8, waxy, slightly compressed, in two fascicles.

A large genus of perhaps 350–400 species, widespread in tropical Asia from India across to Taiwan and south and east through the Malay Archipelago to the Philippines, Australia, the Solomon Islands, New Caledonia, Vanuatu, Fiji, Samoa and Tahiti. Three species have been found in Samoa.

1. Flowers, peduncle and rachis glabrous ··············· **E. rostriflora**
   Flowers, peduncle and rachis pubescent ····················· **2**
2. Lip spathulate, entire; mentum cylindrical-clavate; leaves two
   ·· **E. robusta**
   Lip 3-lobed; mentum rounded; leaves 2–4 ·············· **E. kingii**

E. kingii *F.Muell.*, South Sci. Rec. 2(4): 71 (1882). Type: Solomon Islands, *Goldfinch* s.n.(holo. MEL!).

Fig. 9. *Oberonia heliophila.* **A**, habit × ⅔; **B**, flower × 14. *O. bifida.* **C**, habit × ⅔; **D**, flower × 14. **A,B**, drawn from *Cribb & Morrison* 1770; **C, D**, drawn from *Cribb & Morrison* 1748. All drawn by Sue Wickison.

E. dolichocarpa Schltr. in Repert. Spec. Nov. Regni Veg.·9: 105 (1911). Type: Samoa, *Vaupel* 564 (holo. B†), **synon.nov.**
For further synonymy see M.Clements (1989) in Austr. Orchid Research 1: 76.

Medium-sized to large epiphyte with ovoid to cylindrical-conical pseudobulbs, 5–18 cm long, 1.2–2 cm in diameter, 2–4-leaved towards apex. Leaves suberect, oblanceolate to oblong-lanceolate, acute, 8–42 × 1.2–4 cm. Inflorescences suberect, densely many-flowered, 5–25 cm long, densely hairy; bracts lanceolate, subacute, pubescent, 2–7 mm long. Flowers pale yellow with a yellow lip, densely hairy on outside of sepals; pedicel and ovary 10–26 mm long. Dorsal sepal broadly ovate, obtuse, 3–4.5 × 2.5–3 mm. Lateral sepals obliquely triangular, obtuse, 4–5 × 4–4.5 mm; mentum rounded, 3–4 mm long. Petals elliptic, obtuse, 3.5–4 × 2 mm. Lip recurved in middle, 3-lobed in middle, 4–5 × 3–4 mm; side lobes small, obtuse, erect; midlobe circular; callus obscure of three thickenings on veins at base of midlobe. Column short, 1.5 mm long.

DISTRIBUTION: Savai'i, 'Upolu. Also in Moluccas, New Guinea, Solomon Islands and North-east Australia.
HABITAT: Uncommon epiphyte found in lowland and montane forest; 200–600 m.
COLLECTIONS: *Christophersen* 366 (BISH), 416 (BISH); *McKee* 3009 (BISH); *Vaupel* 564 (B†); *Whistler* 70a (HAW), 640 (HAW), 4446 (BISH, HAW), 5144 (HAW), 9370 (HAW).

E. robusta *(Blume) Lindl.*, Gen. Spec. Orch. Pl.: 69 (1830). Type: Java, *Blume* s.n. (holo. L!).
Dendrolirium robustum Blume, Bijdr. Fl. Ned. Ind.: 347 (1825).
Eria aeridostachya Rchb.f. ex Lindl. in J. Proc. Linn. Soc., Bot. 3: 48 (1859).
    Type: Fiji, *Seeman* 609 (holo.W!).

A large erect epiphyte 30–60 cm tall. Pseudobulbs clustered, ovoid to oblong-ovoid, 4.5–7 cm long, c. 2 cm in diameter, covered by fleshy cataphylls when young, bifoliate in apical part. Leaves erect, lanceolate, acute, 25–45 × 2.5–5.5 cm. Inflorescence axillary from upper nodes, erect, unbranched, 15–30 cm long, densely many-flowered, covered all over by dense stellate hairs. Flowers white to creamy yellow; pedicel and ovary 1.5–2.5 cm long. Dorsal sepal oblong to oblong-ovate, subacute, 3.5–4 × 2–2.3 mm. Lateral sepals obliquely ovate, acute, 3.5–4 × 3–3.5 mm; mentum clavate, up to 4 mm long. Petals falcate-oblong, obtuse, 2.7–3.2 × 1–1.2 mm. Lip erect, recurved towards apex, oblong-ovate, broadly acute at apex, 4–5 × 2.5–3 mm, apical margins undulate. Column 1 mm long; foot 3.5–4 mm long.

DISTRIBUTION: Savai'i, Tutuila, 'Upolu. Also in India, throughout Asia, into parts of Malesia (Sumatra, Java, the Philippines) and eastwards to the Solomon Islands, Vanuatu, New Caledonia and Fiji.

HABITAT: Epiphyte found in lowland and montane forest and montane lava fields; 300–1550 m.

COLLECTIONS: *Rechinger* 1079 (W); *Whistler* 486 (HAW), 1266 (HAW), 2490a (BISH, HAW, K), 3504 (HAW), 3990 (BISH, HAW), 5149 (HAW), 6800 (HAW), 7019 (HAW), 7040 (HAW), 8135 (HAW), 8722 (HAW), 8936 (HAW), 9381 (HAW), 9400 (HAW); *Wisner* 145 (BISH).

E. rostriflora *Rchb.f.* in Seem., Fl. Vit.: 301 (1868). Type: Fiji, Viti Levu, *Seemann* 615 (holo. W!, iso. AMES!, BM!, K!, P!).

E. vieillardii Rchb.f. in Linnaea 41: 86 (1877).Type: New Caledonia, *Vieillard* 1335 (holo. P!, part of holo. W!).

E. curvipes Kraenzl. in Notizbl. Bot. Gart. Berlin-Dahlem 5: 110 (1909). Type: Samoa, *Vaupel* 416 (holo. B †, iso. AMES!).

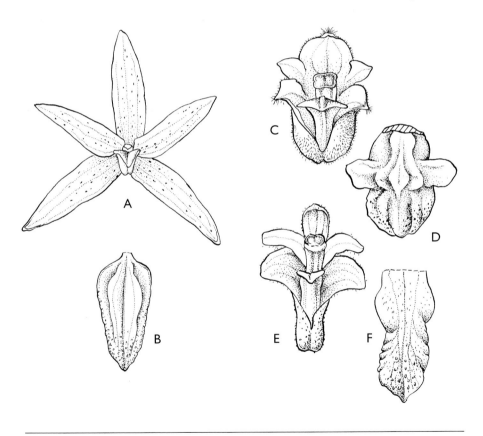

Fig. 10. *Eria rostriflora.* **A**, flower × 4; **B**, lip × 6. *E. kingii.* **C**, flower × 4; **D**, lip × 6. *E. robusta.* **E**, flower × 4; **F**, lip × 6. **A,B** drawn from *Smith & Newmarch* 1/006; **C,D** from *Cribb & Morrison* 1845; **E,F** from *Wickison* 156. All drawn by Sue Wickison.

E. consimilis H.Fleischm. & Rech. in Denkschr. Kaiserl. Akad. Wiss., Math.-Naturwiss. Kl. 85: 206 (1910).Type: Samoa, *Rechinger* 96 (holo. W!).

E. setchellii Schltr. ex Setchell in Univ. Calif. Publ. Bot.: 162 (1926). Type: see *E. vieillardii.*

E. sp. sensu Christophersen in Bernice P. Bishop Mus. Bull. 128: 67 (1935).

An erect epiphyte 15–35 cm tall. Pseudobulbs clustered, erect, cylindrical, 4–25 cm long, 1–1.5 cm in diameter, leafy in apical part. Leaves distichous, ascending, linear-lanceolate or linear-ligulate, acute, 6–20 × 1–2 cm. Inflorescences 1–several, ascending, 5–20 cm long, laxly few to many-flowered, glabrous or sparsely hairy; peduncle short, less than 1 cm long. Flowers pale creamy yellow to pale green, lip marked with a dark purple side lobes; pedicel and ovary 6–10 mm long. Dorsal sepal lanceolate, acute, 4–10 × 0.7–1.5 mm. Lateral sepals obliquely lanceolate, acute, 4–10 × 1.5–2.5 mm. Petals lanceolate, acute, 3.5–9.5 × 0.7–1.5 mm. Lip porrect, slightly curved, ovate to oblong-ovate, entire, acute to acuminate, 2–4 × 0.7–1.5 mm, the disc naked or with two small keels near the base. Column 0.5–1 mm long; foot perpendicular to the column and 0.5–1 mm long.

DISTRIBUTION: Savai'i and 'Upolu. Also in the Mariana Islands (Guam), the Solomon Islands, Vanuatu, Fiji and Tahiti.

HABITAT: Epiphyte found in montane and cloud forests; 600–1780 m.

COLLECTIONS: *Christophersen* 17 (BISH), 365 (BISH), 417 (BISH); *Rechinger* 96 (W); *Vaupel* 416 (AMES, B†); *Whistler* 70 (HAW), 282 (HAW), 708 (HAW), 2535 (HAW, K), 8299 (HAW), 8367 (HAW), 8877 (HAW).

27. **MEDIOCALCAR**

J.J.Sm. in Bull. Inst. Bot. Buitenzorg 7: 3 (1900)

Small epiphytic herbs with creeping rhizomes and small short pseudobulbs. Pseudobulbs clustered to well-spaced, 1–5-leaved, covered when young by cataphylls. Leaves spreading to erect, coriaceous to fleshy. Inflorescences usually one-flowered, solitary or rarely fasciculate, terminal on new shoot. Flowers campanulate, orange, yellow or red, often with yellow or green tips to the sepals and petals. Sepals fused in basal part; the laterals distinctly saccate at base. Petals free but smaller than sepals. Lip with a short broad claw, entire, saccate or spurred, ovate, acute or acuminate. Column relatively long and slender, with a foot; anther terminal; pollinia 8, waxy, in two fascicles of four.

A genus of about 35–40 species centred on New Guinea but extending to the Solomon Islands, Vanuatu, Micronesia, Fiji and Samoa where a single species has been recorded.

M. parodoxum *(Kraenzl.)Schltr.* in Repert. Spec. Nov. Regni Veg. 9: 96 (1910).Type: Samoa, *Reinecke* 300 (holo. B!, part of holo. HBG!, iso. G!).
Eria paradoxa Kraenzl. in Bot. Jahrb. Syst. 25: 606 (1898).

Mediocalcar vanikorense Ames in J. Arnold Arb. 13: 136(1932). Type: Santa Cruz Islands, Vanikoro, *Kajewski* 641 (holo. AMES!).
Bulbophyllum sp. indet. p.p. sensu Setchell, American Samoa 1: 102 (1924).

A creeping epiphyte with elongate rhizomes, 2.5–4 mm in diameter, covered by overlapping cataphylls. Pseudobulbs 2–4.5 cm apart, obliquely ovoid-conical, 0.7–1.6 cm long, 0.4–0.8 cm in diameter, uni- or rarely bifoliate at apex. Leaf oblong to oblong-lanceolate, acute, 4–13 × 0.8–2.3 cm. Inflorescences 1–2-flowered from new growth. Flowers orange or yellow; pedicel slender, elongate, 1.5–3.5 cm long. Sepals connate on basal two-thirds. Dorsal sepal oblong-ovate, acute, 7–8.5 × 3–3.5 mm. Lateral sepals obliquely ovate, acute, 7.5–9 × 4.5–5.5 mm. Petals narrowly oblanceolate, acute, 6.5–8 × 1.5–2 mm. Lip with an oblong broad claw, saccate, ovate, acute, recurved, 6–6.5 × 3.5–4.5 mm. Column c. 4 mm long; foot about 1.5 mm long.

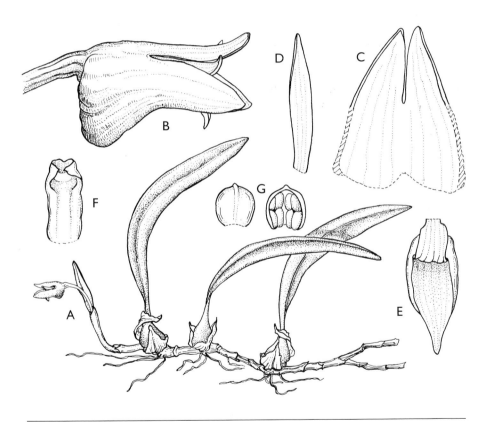

Fig. 11. *Mediocalcar paradoxum*. **A**, habit × ⅔; **B**, flower × 4; **C**, dorsal sepal & lateral sepal × 4; **D**, petal × 4; **E**, lip × 4; **F**, column × 4; **G**, anther cap × 6. **A** drawn from *Wickison* 22: **B–G** from *Mitchell* 9. All drawn by Sue Wickison.

DISTRIBUTION: Tutuila and 'Upolu. Also in the Solomon Islands, the Santa Cruz Islands, Vanuatu and Fiji.

HABITAT: Creeping epiphytic orchid found in lowland to montane forests, and montane scrub; sea level–700 m.

COLLECTIONS: *Christophersen* 369 (BISH), 3488 (AMES, BISH); *Reinecke* 300 (B†, G, HBG); *Setchell* 402 (UC); *Whistler* 713 (BISH, HAW), 1622 (HAW), 3853 (BISH, HAW), 4140 (BISH,HAW), 5715 (HAW), 9187 (HAW), 9621 (HAW).

28. EPIBLASTUS

Schltr. in K.Schum. & Lauterb., Nachtr. Fl. Schutzgeb. Sudsee; 136 (1905)

Large epiphytic or rarely terrestrial herbs with ascending superposed sympodial pseudobulbs; roots fleshy, hairy. Pseudobulbs cylindrical, homoblastic, unifoliate, partially sheathed by persistent cataphylls. Leaves linear to linear-lanceolate, coriaceous. Inflorescences terminal, axillary, fasciculate, one-flowered; peduncles elongate, slender. Flowers not opening widely, often pink or red. Dorsal sepal free. Lateral sepals larger, oblique, forming a mentum with the column-foot. Petals free, smaller than the sepals. Lip attached to the column-foot by a broad crest, entire or weakly 3-lobed with a prominent callus on disc, rarely naked. Column short, stout with a prominent foot; pollinia 8, waxy.

A small genus of about 20 species centred on New Guinea but extending into the Pacific Islands. A single species is reported from Samoa.

E. sciadanthus *(F.Muell.)Schltr.* in K.Schum. & Lauterb., Nachtr. Fl. Schutzgeb. Sudsee: 137, in adnot.(1905). Type: Samoa, *Betche* s.n. (holo. MEL!).
Bulbophyllum sciadanthum F.Muell. in South Sci. Rec. 2: 95 (1882).
Eria ornithidioides Kraenzl. in Bot. Jahrb. Syst. 25: 606 (1898).Type: Samoa, *Reinecke* 313 (holo. B†, iso.G!, without flowers, illustration W!).
Epiblastus ornithidioides (Kraenzl.) Schltr. in K.Schum & Lauterb., Nachtr. Fl. Schutzgeb. Sudsee: 137 (1905). Types: New Guinea, *Schlechter* 14170, 14487 & 14670 (all syn. B†).
Eria sciadantha (F.Muell.)Kraenzl. in Engler, Pflanzenr. Orch. Mon. Dendr. 2: 27 (1911).

Epiphytic scandent plant up to 60 cm long. Pseudobulbs cylindrical, somewhat dilated and compressed above, 5–9 cm long, 1–1.5 cm in diameter, unifoliate. Leaf erect, ligulate to linear-lanceolate, subacute, 25–36 × 1.3–3 cm. Inflorescences fasciculate, 4–10; bracts erect, minute, attenuate-subulate, c. 1 mm long. Flowers campanulate, salmon-pink to reddish pink with whitish tips, glabrous; pedicels slender. Dorsal sepal erect, oblong-elliptic to ovate, acute, 7–8 × 3–4 mm. Lateral sepals obliquely ovate, acute to subacute, 7–8 × 5 mm, saccate at base on lower margin, connate with the column-foot forming a prominent saccate mentum. Petals oblong-elliptic, subacute, 6.5–7.5 × 3–3.5 mm. Lip attached to the apex of the column-foot by a broad claw, ovate, acute, 4.5–6 × 3.5–4.5 mm, the side margins erect; callus basal, transverse, plate-like, deeply medially cleft, weakly undulate in front. Column 3 mm long; foot 4–5 mm long.

DISTRIBUTION: Ta'u, Savai'i and 'Upolu. Also in the Solomon Islands, Vanuatu and Fiji.

HABITAT: Uncommon epiphyte found in lowland and montane forests; 600–1550 m.

COLLECTIONS: *Betche* s.n. (MEL); *Reinecke* 313 (B†, G, W); *Whistler* 1108 (BISH, HAW), 1184 (BISH, HAW), 2023 (HAW), 2555 (BISH, HAW, K), 3693 (BISH, HAW), 5140 (HAW), 7950 (HAW), 8258 (HAW), 9514 (HAW).

29. **AGROSTOPHYLLUM**
Blume, Bijdr. Fl. Ned. Ind.: 368 (1825)

Medium-sized to large epiphytic plants with short rhizomes and clustered stems. Stems often bilaterally flattened, elongate, jointed, leafy, covered by persistent leaf sheaths. Leaves distichous, coriaceous, oblong-ligulate to linear-ligulate, articulated to persistent leaf bases. Inflorescences terminal or rarely lateral, usually dense, comprising many 1–6-flowered spikes in a nodding head, rarely fasciculate, elongate and branching. Flowers small, non-resupinate. Sepals free, spreading. Petals free, smaller than sepals. Lip adnate to base of column, immobile, entire, saccate or spurred at base, with a transverse callus in apical part. Column short to relatively long, with an indistinct foot; pollinia 8, waxy, attached to a common viscidium.

A genus of about 60 species found in S.E. Asia, the Malay Archipelago, Seychelles, New Guinea, the Philippines, Micronesia, New Caledonia and S.W. Pacific Islands east to Samoa. A single species reported from Samoa.

A. megalurum *Rchb.f.* in Seem., Fl. Vit.: 296 (1868). Type: Samoa, Upolu, *Graeffe* s.n.(holo.W).

A large epiphyte up to 75 cm tall. Stems slender, unbranched, 2.5–8 mm across. Leaves ascending, linear-ligulate, shortly and acutely bilobed at apex, 5.3–9.5 × 0.6–1.3 cm; basal sheath oblong-elliptic, 2.2–2.8 cm long, with a small subdeltoid tooth on each side at apex. Inflorescence apical, elongate, fasciculate, 15–28 cm long, racemose, each subtended by a large cataphyll; bracts broadly ovate, clasping, 4.5–5.5 mm long. Flowers white, small. Dorsal sepal oblong to oblong-ovate, acute, 3.5–4 × 1.7–2 mm. Lateral sepals ovate, obliquely oblong-ovate, acute, 4–4.5 × 1.5–1.8 mm. Petals elliptic-ovate to narrowly ovate, subacute, 4 × 1.3 mm. Lip closely appressed to column, oblong-pandurate, abruptly acuminate at apex, 3–3.5 × 1.5 mm, somewhat saccate at base; callus a transverse crest in middle of lip. Column c. 2.5 mm long.

DISTRIBUTION: Savai'i, Ta'u and 'Upolu. Also found in Fiji.

HABITAT: Uncommon epiphyte in montane forest; 200–600 m.

COLLECTIONS: *Betche* s.n. (BISH); *Graeffe* 1259 (BISH), s.n. (BISH, W); *Reinecke* 297 (B); *Vaupel* 547 (B†); *Whistler* 21 (BISH, HAW), 5366 (BISH, HAW), 6949 (BISH, HAW), 8324 (BISH, HAW), 9535 (HAW); *Yuncker* 9246 (BISH).

30. APPENDICULA

Blume, Bijdr. Fl. Ned. Ind.: 297 (1825)

Small to large epiphytic or lithophytic plants with clustered stems on a short rhizome. Stems unbranched or rarely laxly branched near base, leafy, covered by persistent sheathing leaf-bases. Leaves distichous, relatively thin-textured, ligulate to narrowly elliptic, often twisted at base to lie in one plane. Inflorescences terminal or axillary, simple to branched, short to long, few to many-flowered. Flowers small, resupinate, white, greenish or yellow. Dorsal sepal free, erect. Lateral sepals united at base to column-foot to form a more or less saccate mentum. Petals free, smaller than sepals. Lip adnate to apex of column-foot, immobile, more or less entire or weakly 3-lobed, with a more or less horseshoe shaped transverse callus. Column short with a prominent foot; pollinia 6, waxy, clavate to pear-shaped.

A genus of about 100 species widespread in tropical Asia and S.E. Asia, the Malay Archipelago, New Guinea, the Philippines and the S.W. Pacific Islands across to New Caledonia and Samoa. A single species has been found in Samoa. Kores (1989) reported a second species, *A. reflexa* Blume from Samoa but we have not been able to confirm this.

A. bracteosa *Rchb.f.* in Seem., Fl. Vit.: 299 (1868). Type: Fiji, Viti Levu, *Seemann* 592 (syn. W!); Samoa, Upolu, *Graeffe* 11 (syn. W!).

Lobogyne bracteosa (Rchb.f.) Schltr. in Mém. Herb. Boissier 21: 65 (1900).

A. pendula sensu Kraenzl. in Bot. Jahrb. Syst. 25: 601 (1898); H.Fleischm. & Rech. in Denkschr. Kaiserl. Akad. Wiss., Math.-Naturwiss. Kl. 85: 257 (1910), non Blume.

An erect epiphyte. Stems clustered, slender, unbranched, 30–50 cm long. Leaves distichous, narrowly elliptic to elliptic-lanceolate, obtuse, 2.5–5 × 0.6–2.5 cm; sheaths tubular, 0.6–1.1 cm long. Inflorescences terminal and rarely axillary, simple or rarely branching, pendulous, laxly many-flowered; bracts reflexed with age, ovate to ovate-elliptic, 2–6 mm long. Flowers pale green or greenish white. Dorsal sepal ovate, attenuate, 3–3.5 × 1–1.5 mm. Lateral sepals similar but oblique. Petals oblong, obtuse, 2.8–3.2 × 1–1.5 mm. Lip sessile, oblong to subpandurate, obtuse, 2.8–3 × 1.3–2 mm. Column 1.5 mm long.

DISTRIBUTION: Olosega, Savai'i, Ta'u, Tutuila, 'Upolu. Also found in the Solomon Islands, Vanuatu and Fiji.

HABITAT: Epiphyte in montane and cloud forest; 300–1200 m.

COLLECTIONS: *Christophersen* 68 (BISH, K), 1138 (BISH), 3160 (BISH), 3238 (BISH); *Cox* 171a (BISH); *Garber* 714 (BISH), 730 (BISH); *Graeffe* 11 (W), 1260 (BISH, HBG); *Johansson* 8075 (S); *Powell* 270 (K); *Rechinger* 82 (W), 426 (W), 1190; *Reinecke* 307 (B†), 307a (B†), 370 (B†); *Sledge* 1678 (K); *Vaupel* 284 (AMES, K); *Whistler* 356 (BISH, HAW), 535 (HAW), 786 (HAW), 1397 (HAW), 3092 (BISH, HAW), 3170 (BISH, HAW, K), 3820 (HAW, K), 4773 (HAW), 6946 (HAW), 7025 (HAW), 7804 (HAW), 7951 (HAW), 7999 (HAW), 8740 (HAW); *Yuncker* 9267 (BISH).

31. **PHREATIA**

Lindl., Gen. Spec. Orch. Pl.: 63 (1830)

Dwarf to medium-sized epiphytic herbs. Stems short to long or pseudobulbous, leafy. Leaves solitary to many, somewhat thin-textured to fleshy or leathery, articulate. Inflorescences basal or axillary, unbranched, racemose, laxly to densely many-flowered. Flowers resupinate, small or tiny, white, rarely pale green or pale yellow. Dorsal sepal free. Lateral sepals oblique, attached at base to column-foot to form a distinct mentum. Petals narrower than dorsal sepal, free. Lip adnate to column-foot, immobile, concave or saccate at base, usually with two small pit-like nectaries, the disc naked or papillate. Column short, with a foot; pollinia 8, borne on a small linear stipe attached to a small viscidium.

A genus of 100 or more species widespread in tropical Asia from India and Sri Lanka across to New Guinea, N.E. Australia and the S.W. Pacific Islands. Four species in Samoa.

1. Stem pseudobulbous, bifoliate · **P. paleata**
   Stem not pseudobulbous, very short to cylindrical, bearing several leaves in a fan · **2**
2. Plant 1.5–3 cm tall; leaves fleshy, less than 3 cm long · · · · · · · **P. minima**
   Plant more than 5 cm tall, often much larger · · · · · · · · · · · · · · · · **3**
3. Leaves 11–30 × 1.2–2 cm · **P. micrantha**
   Leaves 4–10.5 × 0.2–0.5 cm · **P. myosurus**

P. micrantha *(A.Rich.)Schltr.* in Repert. Spec. Nov. Regni Veg. 1: 919 (1913). Type: Santa Cruz Islands, Vanikoro, *Lesson* s.n. (holo. P!).

Oberonia micrantha A.Rich. in Sert. Astrol.: 7, tab. 3 (1833).

Eria sphaerocarpa Rchb.f. in Seeman, Fl. Vit. 301 (1868). Type: Fiji, *Graeffe* s.n. (holo. W!).

Thelasis samoensis Kraenzl. in Bot. Jahrb. Syst. 25: 607 (1898). Type: Le Pua [?Lepué], *Reinecke* 606 (syn. B†, isosyn. AMES!)

Phreatia macrophylla Schltr. in Bot. Jahrb. Syst. 39: 78 (1906). Type: New Caledonia, *Schlechter* 15465 (holo. B†, iso. BM!, K!, P!, W!, Z!).

P. samoensis (Kraenzl.)Schltr. in Repert. Spec. Nov. Regni Veg. 3: 320 (1907).

P. sarcothece Schltr. in Repert. Spec. Nov. Regni Veg. 9: 438 (1911). Type: Vanuatu, *Morrison* s.n. (holo. B†).

P. graeffei Kraenzl. in Engler, Pflanzenr. Orch. Thelas.: 26 (1911). Based on same type as *Eria sphaerocarpa* Rchb.f.

P. collina Schltr. in Repert. Spec. Nov. Regni Veg. Beih. 1: 919 (1913).Type: New Guinea, *Schlechter* 16438 (holo. B†); non J. J. Smith (1911).

P. robusta R.Rogers in Trans. Roy. Soc. Austr. 54: 39 (1930).Type: Australia, *A.Beck* s.n. (holo. AD).

Rhynochophreatia micrantha (A.Rich.)N.Hallé in Fl. Nouv. Caléd. 8: 341 (1977).

Medium-sized epiphyte with a very short stem. Leaves arranged in a fan, 5–10, ligulate, unequally obtusely bilobed, 11–30 × 1.2–2.2 cm, articulated to broad conduplicate imbricate leaf bases, 3–5 mm long. Inflorescences axillary, usually slightly longer than the leaves, cylindrical, subdensely many-flowered; peduncle twice as long as the rachis; bracts ovate, subulate, 2.5–4 mm long. Flowers tiny, white; pedicel and ovary 1.5–2 mm long. Dorsal sepal broadly ovate, obtuse, 1.2–1.5 × 1–1.5 mm. Lateral sepals oblique, broadly ovate, obtuse, 1–1.5 × 0.9–1.2 mm; mentum short, obtuse. Petals broadly ovate, subacute, 0.8–1.1 × 0.7–1 mm. Lip inflexed, subrhombic to broadly trullate, obtuse, 1–1.7 × 1.3–1.6 mm, lacking a spur. Column 0.4 mm long; foot 0.7 mm long.

DISTRIBUTION: Olosega, Savai'i, Ta'u, Tutuila and 'Upolu. Also found from Micronesia, N. Australia, New Guinea, the Solomon Islands, Vanuatu and New Caledonia eastwards to Samoa.

HABITAT: Epiphyte found in coastal and montane forests; sea level to 1030 m.

COLLECTIONS: *Bristol* 1985 (BISH); *Christophersen* 130 (BISH), 173 (BISH), 627 (BISH), 961 (BISH), 1133 (BISH), 1189 (BISH), 2187 (BISH), 2318 (BISH), 2920 (BISH), 3503 (BISH), 3538 (BISH); *Cox* 340 (BISH); *Garber* 745 (BISH); *Graeffe* 1249 (HAW); *Kennedy* 3985 (BISH); *Long* 3018 (HAW); *Mitchell* 581 (BISH); *Rechinger* 437 (W), 931 (W), 957 (W); *Reinecke* 606 (AMES, B†), 652 (B†); *Setchell* 400 (UC); *Sledge* 1711 (K); *Vaupel* 155 (AK, B, BISH, HAW); *Whistler* 22 (BISH, HAW), 158 (BISH, HAW), 488 (HAW), 2740 (BISH, HAW), 3160 (HAW), 4709 (HAW), 5150 (HAW), 6869 (HAW), 8003 (HAW), 9002 (HAW); *Whitmee* 244 (K).

P. minima *Schltr.* in Repert. Spec. Nov. Regni Veg. Beih. 1: 922 (1913). Type: Papua New Guinea, Dischore Mts., *Schlechter* 19734 (holo. B†).

Tiny epiphyte, 1.5–3 cm tall, with a short stem. Leaves 3–5, fleshy, linear-oblanceolate, obtuse, 1–3 × 0.2–0.3 cm. Inflorescences lateral, as long as or shorter than the leaves, laxly several-flowered; bracts minute. Flowers not opening widely, translucent white or pale green, probably self-pollinating. Dorsal sepal ovate, subacute, 1 × 0.6–0.7 mm. Lateral sepals obliquely ovate, subacute, 1 × 0.7–0.8 mm; mentum shortly conical. Petals narrowly oblong-elliptic, obtuse, 0.7 × 0.3 mm. Lip obovate-spatulate, rounded at apex, 0.6–0.7 × 0.5 mm; claw oblong. Column very short; foot 0.5 mm long.

DISTRIBUTION: Savai'i and 'Upolu. Also in New Guinea and the Solomon Islands.

HABITAT: Epiphyte in montane forest; 575–670 m.

COLLECTIONS: *Whistler* 165 (HAW), 2047a (HAW), 8292 (HAW), 9364 (HAW).

P. myosurus *(G.Forst.)Ames*, Orchidaceae 2: 203 (1908). Type: Society Islands, *Banks & Solander* s.n.(holo. BM!).

Eria myosurus (G.Forst.)Rchb.f. in Bonplandia 5: 54 (1857).

E. stachyurus Rchb.f. in Seem., Fl. Vit.: 301 (1868).

Phreatia matthewsii Rchb.f. in Otia Bot. Hamb.: 55 (1878). Type: Society Islands, *Matthews* s.n. (holo. W!)

P. minutiflora sensu Kraenzl. in Bot. Jahrb. Syst. 25: 607 (1898), non Lindl.

Thelasis sp. sensu Kraenzl. *loc. cit.*

Phreatia inversa Schltr. in K.Schum. & Lauterb., Nachtr. Fl. Deutsch. Sudsee: 187 (1905). Type: New Ireland, *Schlechter* 14644 (holo. B †, iso. BO!, K!).

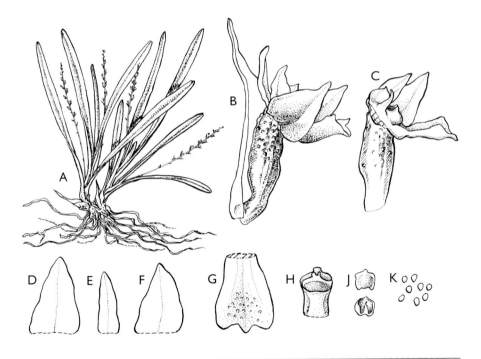

Fig. 12. *Phreatia myosurus.* **A**, habit × ⅔; **B**, flower and bract × 20; **C**, flower with dorsal and nearside sepal and petal removed × 20; **D**, dorsal sepal × 20; **E**, petal × 20; **F**, lateral sepal × 20; **G**, lip × 20; **H**, column × 20; **J**, anther cap two views × 20; **K**, pollinia × 20. All drawn from *Wickison* 128 by Sue Wickison.

P. upoluensis Schltr. in Repert. Spec. Nov. Regni Veg. 3: 319 (1907). Type: Samoa, *Betche* 36 (holo. B†, iso. MEL!).

P. reineckei Schltr. in Repert. Spec. Nov. Regni Veg. 9: 109 (1910). Type: Samoa, *Vaupel* 658 (holo. B†, iso. K!).

P. neocaledonica Schltr. in Bot. Jahrb. Syst. 39: 78 (1906). Type: New Caledonia, *Schlechter* 14755 & 15228 (syn. B†, isosyn. BO!, BM!, K!, P!, Z!).

A small epiphyte with a short stem, 1–4 cm long. Leaves 6–10, arranged in a fan, linear-ligulate, briefly obliquely unequally bilobed at apex, 4–10.5 × 2–5 mm, articulated to sheathing, imbricate leaf-bases. Inflorescences axillary, more or less as long as the leaves, 3–11 cm long, laxly many-flowered; peduncle about half length of rachis; bracts lanceolate, acuminate, 1.5–2 mm long. Flowers tiny, white; pedicel and ovary c. 1 mm long. Dorsal sepal ovate, subacute, 0.8–1.2 × 0.6–0.8 mm. Lateral sepals obliquely ovate, acute, 1–1.5 × 0.9–1.2 mm; mentum subconical. Petals elliptic-ovate, acute, 0.8–1.2 × 0.6–0.8 mm. Lip clawed, with a transversely ovate, obtuse lamina, 0.9–1.3 × 0.7–0.9 mm, lacking a spur. Column short; foot 0.3 mm long.

DISTRIBUTION: Apolima, Ofu, Olosega, Savai'i, Ta'u, Tutuila, 'Upolu. Also in New Ireland, Bougainville, the Solomon Islands, Vanuatu, the Horn Islands, Fiji and the Society Islands.

HABITAT: Epiphyte found in coastal and montane forests; sea level–800 m.

COLLECTIONS: *Betche* 36 (MEL); *Christophersen* 5 (BISH), 41 (BISH), 1057 (AMES, BISH), 1193 (BISH), 1211 (BISH), 1259 (BISH), 1821 (BISH), 2296a (BISH), 3207 (BISH), 3278 (BISH), 3573 (BISH); *Funk* 9 (K); *Garber* 629 (BISH), 938 (BISH); *Graeffe* 1252 (HAW); *McKee* 3007 (BISH); *Powell* 364 (K); *Rechinger* 124 (W), 521 (W), 1497 (W), 1660 (W), 1685 (W), 1697 (W), 1704 (W); *Reinecke* 183 (K), 239 (B†), 292 (B†), 587 (B†); *Sledge* 1708 (K); *Vaupel* 156 (K), 658 (B†, K); *Whistler* 24 (BISH, HAW), 165 (HAW), 505 (HAW), 601 (HAW), 949 (HAW, K), 1593 (HAW), 2047 (HAW), 2698 (HAW, K), 3074 (HAW), 3209 (HAW, K), 4494 (HAW), 5147 p.p. (HAW), 5344 (BM, HAW, K), 6859 (HAW), 6868 (HAW), 7056 (HAW), 7716 (HAW), 7747 (HAW), 8149 (HAW), 8156 (HAW), 8293 (HAW), 8793 (HAW), 9001 (HAW), 9104 (HAW), 9243 (HAW), 9361 (HAW); *Whitmee* 47 (K); *Wisner* 140 (BISH); *Yuncker* 9000 (BISH), 9305 (BISH).

NOTE: Close to and possibly conspecific with *P. stenostachya* (Rchb.f.) Kraenzl. but material of the latter appears to have inflorescences much longer than the leaves and slightly longer pedicels. Their relationship needs further investigation.

P. paleata *Rchb.f.* in Linnaea 41: 653 (1877). Type: New Caledonia, *Vieillard* s.n. (holo. P!).

Eria paleata Rchb.f. in Linnaea 41: 87 (1877). Type: New Caledonia, *Vieillard* 1331 (holo. P!).

P. obtusa Schltr. in Repert. Spec. Nov. Regni Veg. 9: 108 (1910). Type: Samoa, *Vaupel* 527 (holo. B †, iso. AMES!).

P. pholidotoides Kraenzl. in Not. Syst. 4: 140 (1928). Type: New Caledonia, *Cribs* 1217 (holo. P!).

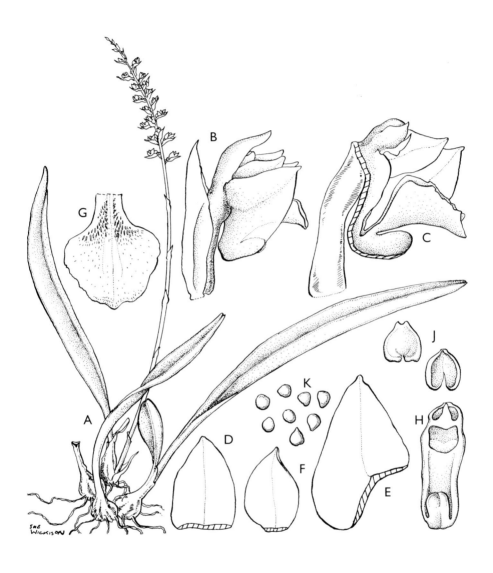

Fig. 13. *Phreatia paleata*. **A**, habit × ⅔; **B**, flower, ovary and bract × 8; **C**, column, lip, spur and ovary × 10; **D**, dorsal sepal × 10; **E**, lateral sepal × 10; **F**, petal × 10; **G**, lip × 10; **H**, column (ventral view) × 10; **J**, anther cap (two views) × 10; **K**, pollinia × 10. All drawn from *Wickison* 17 by Sue Wickison.

Medium-sized epiphyte with a short creeping rhizome. Pseudobulbs proximate, ovoid, 1–2 cm in diameter, bifoliate at apex. Leaves erect, oblong-lanceolate, acute, or subacute, 16–25 × 1.1–2.2 cm, shortly petiolate. Inflorescences basal, lateral, erect, as long as or longer than the leaves, 17–35 cm tall, cylindrical-tapering, densely many-flowered; peduncle 12–25 cm long; bracts cucullate, oblong, obtuse, 2.5–3 mm long. Flowers white, small; pedicel and ovary 3–4 mm long. Dorsal sepal ovate, obtuse, 2.5–3 × 2 mm. Lateral sepals obliquely ovate, subacute, 3 × 1.5–2 mm; mentum short, obtuse. Petals elliptic-ovate, subacute, 1.5–1.8 × 1 mm. Lip shortly clawed, triangular-trullate above, obtuse, 2–2.5 1.5–2 mm, minutely puberulent on disc; spur short, cylindrical, one-third length of lip. Column short; foot well-developed.

DISTRIBUTION: Savai'i, Tutuila and 'Upolu. Also in Vanuatu, New Caledonia and Norfolk Island.

HABITAT: Epiphyte found in rain- and cloud-forests; 400–740 m.

COLLECTIONS: *Funk* 10 (K); *Vaupel* 527 (AMES); *Whistler* 269 (HAW), 709 (HAW, K), 2016 (HAW), 2048 (HAW), 2063 (HAW, K), 6875 (HAW), 7002 (HAW), 9000 (HAW), 9363 (HAW).

NOTE: Close to and perhaps conspecific with *P. tahitiensis* Lindl. from Tahiti. Their relationship needs further investigation.

32. **THELASIS**
Blume, Bijdr. Fl. Ned. Ind. 385, t.75 (1825)

Small to medium-sized epiphytic herbs. Stems pseudobulbous, 1–two-leaved, pseudobulb sometimes subtended by three or four leaves arranged in a fan. Leaves narrow, linear to oblanceolate or oblong, thin-textured or somewhat coriaceous. Inflorescence lateral; peduncle longer than the rachis, slender, several- to many-flowered, racemose. Flowers small, not showy, outcrossing or self-pollinating, sometimes cleistogamous; ovary circular or angular in cross section. Sepals subsimilar, the laterals joined at the base, sometimes keeled. Petals small, more or less equal in length to the petals. Lip entire, widest and excavated at the base, narrowing to tip. Column lacking a foot; rostellum bifid; pollinia eight; stipes slender, spatulate; viscidium lanceolate or elliptic.

A genus of about 20 species widespread from India across tropical Asia to the Malay Archipelago and S.W. Pacific Islands. A single species in Samoa, the easternmost extent of the range of the genus.

T. carinata *Blume,* Bijdr. Fl. Ned. Ind.: 306 (1825). Type: Java, *Blume* s.n. (holo. L, iso. K!).

Oxyanthera papuana Schltr. in K.Schum. & Lauterb., Nachtr. Fl. Deutsch Sudsee: 126 (1905). Type: Papua New Guinea, Torricelli Mts., *Schlechter* 14330 (holo. B†, iso. K!).

A medium-sized epiphyte with clustered growths; pseudobulbs obscure, ovoid, bilaterally flattened, up to 4.5 × 1.5 cm, unifoliate at apex but subtended by several leaves arranged in a fan. Leaves twisted at base to lie in one plane, erect or suberect, linear, unequal bilobed at obtuse apex, channelled below, 12–45 × 1.2–2.5 cm, articulated to a channelled leaf base. Inflorescence erect, 15–35 cm tall; peduncle 12–30 cm long, slender; rachis up to 16 cm long; bracts very closely spaced, reflexed, triangular or lanceolate, 1.5–3 mm long. Flowers small, white or pale yellow, rarely suffused with purple, probably autogamous and perhaps cleistogamous. Dorsal sepal oblong-ovate, acute, 3.5 × 1.5 mm. Lateral sepals oblong-lanceolate, acute, 3–3.5 × 1.5–2 mm, keeled on outer surface. Petals oblong-lanceolate, acute, 2–3 × 0.8 mm. Lip ovate, obscurely 3-lobed in basal part, acute, 3–3.5 × 1.5–2 mm. Ovary pendent, 4 mm long.

DISTRIBUTION: Savai'i and 'Upolu. Widespread from S.E. Asia and the Philippines across to New Guinea and Samoa.

HABITAT: Epiphyte in montane forest; 450–500 m.

COLLECTIONS: *Whistler* 672 (BISH, HAW), 1522 (BISH, HAW), 1681 (BISH, HAW).

33. **EARINA**

Lindl. in Bot. Reg. 20: sub t.1699 (1834)

Large epiphytic or terrestrial herbs with clustered stems arising from short rhizomes. Stems pseudobulbous or elongated and slender, leafy, covered by persistent tubular or imbricate leaf-bases. Leaves distichous, coriaceous, ligulate. Inflorescence terminal, simple or branching, densely many-flowered. Flowers small, resupinate or not, usually white or pale greenish. Dorsal sepal free. Lateral sepals oblique, free. Petals free, narrower than sepals. Lip sessile, immobile, entire or weakly 3-lobed, often saccate at base, channelled above, ecallose. Column long, slender, lacking a foot; pollinia 4, waxy, joined to a small viscidium.

A small genus of about six species in New Zealand, New Caledonia, Vanuatu, the Solomon Islands, Fiji and Samoa. A single species has been recorded from Samoa.

E. **valida** *Rchb.f.* in Linnaea 41: 96 (1876). Type: New Caledonia, *Vieillard* 1298 in part (lecto. P!, illn. & flowers of lecto. W!).

E. *laxior* Rchb.f. in Otia Bot. Hamburg.: 54 (1878) [Repr. Xenia Orchid. 3: 30 (1881)]. Type: Tahiti, *U.S. Expl. Exped.* (this label is doubtful, probably from Fiji or Samoa) (holo. AMES!, iso. W!).

E. *plana* Rchb.f. in Otia Bot. Hamburg.: 54 (1878) [Repr. Xenia Orchid. 3: 30 (1881)]. Type: Fiji, Vanau Levu, *U.S. Expl.Exped.* (holo. AMES!, drawing of holo. and fragments at W!)

E. samoensiana F.Muell. & Kraenzl. in Oesterr. Bot. Zeitschr. 44: 211 (1894). Type: Samoa, *Betche* 55 (holo. MEL!).

Agrostophyllum drakeanum Kraenzl. in J. Bot. 17: 442 (1903).Type: New Caledonia, *Baudouin* 347 (holo. P!).

Earina brousmichei Kraenzl. in Notul. Syst. 4: 136 (1928). Type: New Caledonia, *Brousmiche* 987 (holo. P!).

An erect epiphyte with clustered pseudobulbs, 4–12 cm long. Leaves suberect to erect, ligulate, unequally obtusely bilobed at apex, 30–90 × 0.8–2.5 cm, articulated to inflated leaf-bases. Inflorescence 50–120 cm long, branching, densely many-flowered; peduncle bilaterally flattened; branches short, distichous, ascending, 3–5-flowered, 0.5–1.2 cm long; bracts broadly ovate, 1–1.5 mm long. Flowers white, small; ovary 4.5–6 mm long. Dorsal sepal oblong-obovate, subacute, 4–4.5 mm long. Lateral sepals obliquely ovate, acute, 4–5 × 3 mm. Petals oblong to elliptic-oblong, subacute, 4.5 × 2.5 mm. Lip entire, oblong-obovate, subacute, 5 × 3 mm, somewhat medially constricted, reflexed at apex, saccate at base. Column 4.5–5 mm long.

DISTRIBUTION: Savai'i and 'Upolu. Also in Vanuatu, New Caledonia and Fiji.
HABITAT: Epiphyte found in montane forest; 500–900 m.
COLLECTIONS: *Betche* 55 (MEL); *Christophersen* 2911 (BISH, P); *Powell* 249 (K); *Reinecke* 614 (B†); *Vaupel* 544 (B†); *Whistler* 903 (HAW), 1258 (HAW), 1919 (HAW), 2013 (HAW).

34. **GLOMERA**
Blume, Bijdr. Fl. Ned. Ind.: 372 (1825)

Erect or pendulous epiphytic herbs with slender elongate leafy stems, often branching, many-noded, covered by persistent leaf-sheaths. Leaves more or less distichous, coriaceous, linear-lanceolate, articulated to tubular leaf bases that can be striate or warty. Inflorescences terminal on branches, 2–many, dense, capitate, enclosed below by imbricate sheathing bracts. Flowers small, non-resupinate. Dorsal sepal free or partially connate with lateral sepals. Lateral sepals oblique, connate in basal part, enclosing base of lip. Petals free, narrower than dorsal sepal. Lip adnate to column-foot, immobile, entire, sometimes fleshy, saccate at base, the epichile separated from the base by a transverse ridge. Column short, with a short foot; pollinia 4, waxy, in two pairs or a single fascicle.

A genus of about 50 species centred on New Guinea but extending to Moluccas and the South West Pacific Islands across to the Carolines, Vanuatu, Samoa and Fiji. A single species reported from Samoa.

G. montana *Rchb.f.* in Linnaea 41: 77 (1876).Type: Fiji, *Milne* s.n. (holo. K!).

Agrostophyllum reineckeanum Kraenzl. in Bot. Jahrb. Syst. 25: 602 (1898). Type: Samoa, 'Upolu, *Reinecke* 297 (holo. B†; part of holo. HBG!; iso. BISH!, E!, G!).

PLATE 1

A. Corymborkis veratrifolia

C. Goodyera rubicunda

B. Erythrodes oxyglossa

D. Moerenhoutia heteromorpha

PLATE 2

A. Pristiglottis longiflora   B. Hetaeria whitmeei
C. Vrydagzynea samoana   D. Vrydagzynea vitiensis

PLATE 3

A. Zeuxine stenophylla
C. Spiranthes sinensis

B. Zeuxine vieillardii
D. Cryptostylis arachnites

PLATE 4

A. Peristylus whistleri
C. Nervilia aragoana

B. Habenaria samoensis
D. Didymoplexis micradenia

PLATE 5

A. Calanthe hololeuca

B. Calanthe triplicata

C. Calanthe alta

PLATE 6

A. Phaius tankervilleae   B. Phaius terrestris
C. Phaius flavus

PLATE 7

A. Calanthe ventilabrum
C. Malaxis reineckiana

B. Spathoglottis plicata

PLATE 8

Coelogyne lycastoides

PLATE 9

A. Malaxis resupinata
C. Malaxis samoensis

B. Malaxis tetraloba
D. Malaxis taurina

PLATE 10

A. Liparis layardii

C. Liparis gibbosa

B. Liparis condylobulbon

D. Mediocalcar paradoxa

PLATE 11

A. Oberonia equitans
C. Epiblastus sciadanthus

B. Oberonia heliophylla
D. Appendicula bracteosa

PLATE 12

A. Eria rostriflora   B. Eria robusta
C. Eria kingii

PLATE 13

A. Phreatia myosurus
C. Phreatia minima

B. Phreatia micrantha
D. Phreatia paleata

PLATE 14

A. Thelasis carinata
C. Glomera montana

B. Earina valida
D. Diplocaulobium fililobum

PLATE 15

A. Dendrobium biflorum
C. Dendrobium catillare

B. Dendrobium calcaratum
D. Dendrobium dactylodes

PLATE 16

Dendrobium mohlianum

PLATE 17

A. Dendrobium goldfinchii
C. Dendrobium reineckei

B. Dendrobium samoense
D. Dendrobium lepidochilum

PLATE 18

A. Dendrobium sladei
C. Dendrobium vagans

B. Dendrobium whistleri
D. Dendrobium comata

PLATE 19

A. Pseuderia ramosa
C. Bulbophyllum longiflorum

B. Bulbophyllum betchei
D. Bulbophyllum longiscapum

PLATE 20

Bulbophyllum distichobulbum

PLATE 21

A. Bulbophyllum pachyanthum
C. Bulbophyllum membranaceum

B. Bulbophyllum ebulbe
D. Bulbophyllum samoanum

PLATE 22

A. Bulbophyllum savaiense   B. Thrixspermum graeffei

C. Bulbophyllum trachyanthum

PLATE 23

A. Schoenorchis micrantha
C. Microtatorchis samoensis

B. Luisia teretifolia
D. Pomatocalpa vaupelii

PLATE 24

A. Taeniophyllum savaiiense   B. Taeniophyllum fasciola
C. Chrysoglossum ornatum

Glomera samoensis Rolfe in Bull. Misc. Inform., Kew 1908: 414 (1908). Type: Samoa, *Funk* 11 (holo. K!).

G. gibbsiae Rolfe in J. Linn. Soc. Bot. 39: 176 (1909). Type: Fiji, Viti Levu, *Gibbs* 817 (holo. BM!, iso. K!).

Agrostophyllum megalurum sensu H.Fleischm. & Rech. in Denkschr. Kaiserl. Akad. Wiss., Math.-Naturwiss. Kl. 85: 257 (1910); non Rchb.f.

Fig. 14. *Glomera montana*. **A**, habit × ⅔; **B**, flower × 3: **C**, dorsal sepal × 4; **D**, petal × 4; **E**, lateral sepal × 4; **F**, lip × 6; **G**, column, lip and ovary × 6; **H**, column × 6; **J**, anther cap × 8; **K**, pollinia × 8. All drawn from *Cribb & Wheatley* 12 by Sue Wickison.

Glomera reineckiana (Kraenzl.) Schltr. in Repert. Spec. Nov. Regni Veg. 9: 96 (1911).

G. *rugulosa* Schltr. in Repert. Spec. Nov. Regni Veg. Beih. 1: 287 (1912). Type: New Guinea, *Schlechter* 18200 (holo. B†).

An epiphyte with clustered erect stems up to 120 cm long. Stems slightly bilaterally flattened, leafy, branching. Leaves erect, lanceolate, subacute, 5.5–12 × 0.5–1.5 cm, articulated to a weakly striate tubular leaf-base. Inflorescences nodding, semiglobose, 6–15-flowered, subtended by hyaline bracts. Flowers white with pink tip to the lip; ovary 0.6–1.1 cm long. Dorsal sepal oblong- to elliptic-ovate, subacute, 6.5–7.5 × 2.5–3.2 cm. Lateral sepals connate at base, obliquely ovate, falcate, acute or acuminate, 7–8 × 4–4.5 mm. Petals oblong- to oblong-elliptic, subacute to obtuse, 6–7.5 × 2.3–3 mm. Lip 4.5–5.5 × 2–2.5 mm, the base subglobose, oblong to oblong-obovate above and broadly rounded at apex. Column 2–2.5 mm long, the foot very short.

DISTRIBUTION: Savai'i, Ta'u, Tutuila and 'Upolu. Also found in New Guinea, Bougainville, the Solomon Islands, Vanuatu and Fiji.

HABITAT: Epiphyte found in montane forests; 250–1200 m.

COLLECTIONS: *Christophersen* 123 (BISH, K), 1040 (BISH, P), 2089 (BISH), 3527 (BISH); *Cox* 328 (BISH); *Funk* 11 (K); *Garber* 748 (BISH); *Graeffe* 1262 (HAW); *Long* 2606 (HAW), 3070 (HAW); *Rechinger* 708 (W); *Reinecke* 297 (BISH, E, G, HBG); *Vaupel* 411 (AMES, K); *Whistler* 166 (BISH, HAW), 283 (HAW), 483 (HAW), 802 (HAW), 3182 (BISH, HAW, K), 3959 (BISH, HAW), 7852 (HAW), 9299 (HAW), 9372 (HAW); *Wilkes* s.n. (W).

35. **DENDROBIUM**

Sw. in Nova Acta Regiae Soc. Sci. Upsal. ser.2, 6: 82 (1799)

Small to large epiphytic, lithophytic or rarely terrestrial herbs. Stems elongate, slender and leafy or swollen, pseudobulbous and leafy towards apex, 1–several-noded. Leaves coriaceous to fleshy, flat, terete or bilaterally flattened, articulated at base; leaf base present or not. Inflorescences lateral or subterminal, axillary, 1–many-flowered, fasciculate, racemose or paniculate. Flowers often showy and relatively large. Dorsal sepal free. Lateral sepals adnate to column-foot forming a distinct mentum. Petals free, usually narrower than the sepals. Lip attached to apex of column foot, mobile or articulated, entire to 3-lobed, lacking a spur, usually with a callus of ridges or appendages. Column short, with a short to long foot; pollinia 4, waxy, lacking an appendage.

A large genus of perhaps 900 to 1000 species widespread in tropical and subtropical Asia, Malesia, the Philippines, Micronesia, N. and E. Australia and the S.W. Pacific across to Tahiti and south to New Caledonia and New Zealand. Fourteen species have been recorded in Samoa. The taxonomy of sect. *Grastidium* is particularly difficult and more field work is necessary in Samoa and adjacent archipelagoes to assess the species and their synonymies.

1. Leaf solitary, pendent, terete · **D. vagans**
   Leaves two or more, flat or bilaterally flattened · · · · · · · · · · · · · · · · 2
2. Flowers borne in pairs in sessile inflorescences along the stems and
   opposite the leaves · 3
   Flowers not borne in sessile inflorescences, usually three- or more-
   flowered · 7
3. Leaves linear-tapering, acuminate, less than 6 mm in width · · · · · · · · 4
   Leaves lanceolate to narrowly ovate, obtuse or subacute, more than 10
   mm in width · 5
4. Flowers white or pale yellowish · **D. biflorum**
   Flowers pale yellow speckled with red or purple · · · · · · · · **D. reineckei**
5. Sepals acute and not fleshy; petals narrowly linear, acute; lip side lobes
   obscure, blunt; flowers white · **D. dactylodes**
   Sepals blunt and fleshy; petals obliquely linear or falcate-lanceolate, acute;
   lip side lobes acute; flowers pale yellow or yellow flecked with red · · · 6
6. Sepals and petal pale yellow; lip with purple side lobes · · · · · · **D. sladei**
   Sepals and petals yellow, finely speckled with reddish purple within; lip
   side lobes reddish flushed · **D. lepidochilum**
7. Flowers covered on outside of ovary and sepals with setose hairs
   · **D. macrophyllum**
   Flowers glabrous on ovary and sepals · 8
8. Leaves bilaterally flattened or subulate; flowers borne on an elongate
   terminal leafless stalk covered by scarious bracts · · · · · · · · · · · · · · 9
   Leaves dorsiventrally flattened; flowers lateral or subterminal · · · · · · 10
9. Leaves bilaterally flattened; lip 3-lobed; callus of 3 lamellae
   · **D. goldfinchii**
   Leaves subulate; lip elliptic, entire; callus of 2 lamellae · · · **D. scirpoides**
10. Mentum elongate, spur-like, longer than the dorsal sepal and petals · 11
    Mentum chin-like, shorter than the dorsal sepal and petals · · · · · · · 13
11. Lip apex upturned and crimped; flowers red or red with a violet lip
    · **D. mohlianum**
    Lip apex not incurved, acute; flowers pale orange, pink or white · · · 12
12. Flowers pale orange · **D. calcaratum**
    Flowers pink or rarely white · **D. catillare**
13. Inflorescences elongate, much longer than the leaves; flowers cream-
    coloured, mustard or dull yellow with purple veins on the lip; leaves
    many in upper part of stem, elliptic, obtuse · · · · · · · · · · · **D. samoense**
    Inflorescences as long as the leaves or shorter; flowers white with purple
    markings on the lip; leaves two, subapical, lanceolate, acute   **D. whistleri**

section RHIZOBIUM

D. vagans *Schltr.* in Repert. Spec. Nov. Regni Veg. 9: 104 (1910). Type: Samoa,
Vaupel 651 (holo. B†, iso. AMES!, K!)
D. crispatum sensu Rchb.f. in Seem., Fl. Vit.: 303 (1868); et auct. non Sw.

D. calamiforme Rolfe in Bull. Misc. Inform., Kew 1921: 55 (1921), *nom illeg.*; non Lodd.(1841).

D. seemannii L.O.Williams in Bot. Mus. Leafl. Harv. Univ. 5: 123(1938).

Dockrillia vagans (Schltr.)Rauschert in Repert. Spec. Nov. Regni Veg. 94: 447 (1983).

Pendent epiphyte up to 200 cm long, with branching wiry rhizome, rooting at base. Stems slender, wiry, cylindrical, 3–6 mm long, unifoliate at apex. Leaf pendent, cylindrical, acute, 6–18 × 0.2–0.3 cm. Inflorescence pendent, racemose, 6–15 cm long, laxly to subdensely 5–10-flowered; peduncle short, 0.7–3 cm long; bracts narrowly ovate, acuminate, 1 mm long. Flowers non-resupinate, white to pale yellow, sometimes tinged with pink; ovary 1.5–2.5 cm long. Dorsal sepal lanceolate, acuminate, 14–20 × 1.5–2 mm. Lateral sepals obliquely lanceolate, acuminate, 14–20 × 2–3.5 mm; mentum 3–5 mm long. Petals linear, acute, 14–20 × 1–1.5 mm. Lip recurved, 3-lobed in middle, 16–20 × 3.5–4.5 mm; side lobes erect, obscure, narrowly oblong; midlobe lanceolate, acuminate, margins strongly undulate, 10–11 mm long; callus of 3 raised undulate keels from base to middle of midlobe. Column 2–2.5 mm long; foot 3–5 mm long.

DISTRIBUTION: Savai'i. Also in Vanuatu and Fiji, possibly extending to New Caledonia and the Loyalty Islands.

HABITAT: Uncommon epiphyte in lowland and montane forest and montane lava fields; sea level to 1600 m.

COLLECTIONS: *Cox* 286 (BISH); *Vaupel* 651 (AMES, K); *Whistler* 2553 (BISH, HAW, K), 4242 (BISH, HAW), 6808 (HAW), 9465 (HAW).

section PEDILONUM

D. calcaratum *A.Rich.* in Sert. Astrol.: 18 (1834). Type: Santa Cruz Islands, Vanikoro, *Lesson* s.n. (holo. P!, iso. W!).

D. triviale Kraenzl. in Bot. Jahrb. Syst. 25: 604 (1898).Type: Samoa, *Reinecke* 422 (holo. B †).

D. separatum Ames in J. Arnold Arb. 13: 133 (1932).Type: Santa Cruz Islands, *Kajewski* 503 (holo. AMES!).

Pedilonum separatum (Ames) Rauschert in Repert. Spec. Nov. Regni Veg. 94(7–8): 463 (1983).

P. triviale (Kraenzl.)Rauschert, loc.cit. 464.

A pendent epiphyte with long clustered ribbed leafy stems up to 1.5 m long, yellow-green turning maroon with age. Leaves deciduous, distichous, coriaceous, ligulate or oblong-lanceolate, acute and unequally bilobed at apex,

---

Fig. 15. *Dendrobium calcaratum*. **A**, habit × ⅔; **B**, flower × 3; **C**, dorsal sepal × 4; **D**, lateral sepal × 4; **E**, petal × 4; **F**, lip × 4; **G**, column and ovary × 6; **H**, anther cap × 6; **J**, pollinia × 6. All drawn from *Wickison* 141 by Sue Wickison.

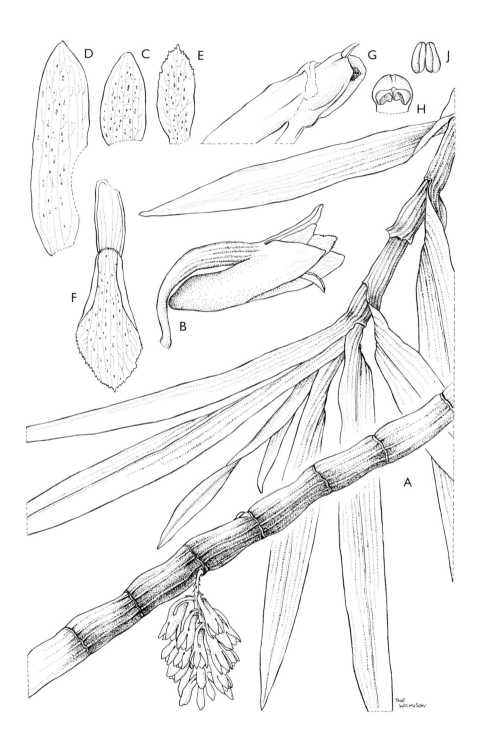

9–16 × 1.3–1.7 cm. Inflorescences several, lateral, pendent, from nodes on leafless stems, densely 10–18-flowered; peduncle up to 2 cm long; bracts 1–3 mm long. Flowers pale apricot to orange, not opening widely; pedicel and ovary c. 1 cm long. Dorsal sepal oblong, obtuse, 5–6 × 3 mm. Lateral sepals obliquely oblong-ovate, obtuse, 14–15 × 3–4 mm; mentum cylindrical, obtuse, 7–8 mm long. Petals elliptic, obtuse, 5–6 × 2–3 mm, with erose margins. Lip obovate, obtuse, 9–10 × 3–4 mm with a lunate transverse callus in basal part. Column 2–3 mm long; foot 8 mm long.

DISTRIBUTION: Savai'i, Tutuila and 'Upolu. Also in New Britain, the Solomon Islands, the Santa Cruz Islands, Vanuatu, Tonga and the Horn Islands.

HABITAT: Mangrove to montane forests, also on lava fields; sea level to 700 m.

COLLECTIONS: *Christophersen* 364 (BISH), 593 (BISH), 3460 (BISH, P); *Graeffe* 1241 (HBG); *Powell* 230 (K, W), 248 (K), 361 (K); *Rechinger* 995 (W); *Reinecke* 422 (B†); *Vaupel* 281 (AMES); *Whistler* 133 (HAW), 480 (HAW), 637 (HAW), 1721 (HAW), 2830 (HAW), 3255 (HAW, K), 3623 (BISH, HAW), 4141 (BISH, HAW), 4572 (HAW), 6791 (HAW), 8708 (HAW); *Wilkes* s.n. (W).

D. catillare *Rchb.f.* in Seem., Fl. Vit.: 304 (1868).Type: Fiji, Kadavu Isl., *Milne in Seeman* 591 (lecto. W!, iso. K!).

D. glomeriflorum Kraenzl. in Gard. Chron. ser. 3, 18: 206 (1895). Type: No type cited.

D. erythroxanthum sensu Kraenzl. in Bot. Jahrb. Syst. 25: 604 (1898), non Rchb.f. For full synonymy see Dauncey & Cribb in Kew Bull. 48: 557 (1993).

A pendent epiphyte with curving many-noded leafy yellow stems, 25–100 cm long. Leaves distichous, somewhat coriaceous, linear-ligulate to lanceolate, subacute to acute, 6.4–16 × 1–2.8 cm. Inflorescences from leafy or leafless stems, capitate, 2–3.5 cm long, 10–25-flowered; peduncle 2–5 mm long; bracts narrowly triangular, acute, 4–10 mm long. Flowers not opening widely, usually white, pink or pale orange; pedicel and ovary, 3-ridged, 9–11.5 mm long. Dorsal sepal lanceolate, subacute, 6–10 × 2–3 mm. Lateral sepals obliquely oblong-lanceolate, subacute, 6–10 × 3–4.2 mm; mentum cylindrical, 6–9 mm long. Petals narrowly elliptic, subacute, 5–10 × 1.5–3 mm. Lip entire or obscurely 3-lobed, more or less ovate, acute or subacute, 3.5–5 × 2–2.6 mm; callus a transverse ridge in basal part. Column short, 1.5–2 mm long; foot 6–8 mm long.

DISTRIBUTION: Savai'i, Ta'u, Tutuila and 'Upolu. Also in Fiji and possibly Tahiti.

HABITAT: Epiphyte in open forest; 250 – 500 m.

COLLECTIONS: *Reinecke* 294 (B†); *Whistler* 751 (HAW), 1631 (BISH, HAW), 2903(BISH, HAW, K), 3183a (HAW), 3238 (HAW), 5154 (HAW), 7985 (HAW), 8945 (HAW), 9362 (HAW), 9521 (HAW); *Wilkes* s.n. (W).

section LATOURIA

D. whistleri *Cribb* in Kew Bull. 50: 785 (1995). Type: Samoa, 'Upolu, *Whistler* 3867 (holo. K!, iso. BISH!, HAW!).
D. punamense auct. non *Schltr.* (1905).

An epiphyte with spreading to pendent subclavate pseudobulbs, 15–22 cm long, 0.5–0.6 cm in diameter above, turning yellow with age, 5–7-noded, 1–2-leaved. Leaves coriaceous, lanceolate, acuminate, 12–18 × 2–2.5 cm. Inflorescences lateral and subterminal, laxly 3–5-flowered, shorter than the leaves; peduncle terete, up to 5 cm long; bracts conduplicate, lanceolate, acuminate, 4–7 mm long. Flowers probably self-pollinating, not opening widely, white with purple on lip; pedicel and ovary 9–15 mm long. Dorsal sepal lanceolate, acuminate, 9–11 × 3–4 mm. Lateral sepals obliquely lanceolate, acuminate, 10–11 × 3.5–4 mm; mentum obliquely conical, 3 mm high. Petals lanceolate, acute, 9–10 mm long, 2 mm wide. Lip longly clawed, 3-lobed, 9 × 6 mm when flattened; side lobes erect, obliquely oblong, somewhat rounded in front; midlobe subquadrate-transversely elliptic, shortly apiculate, 3.5 × 4 mm; callus 3-ridged from base to base of midlobe. Column 3 mm long with an erose apical margin; foot 3 mm long.

DISTRIBUTION: 'Upolu only. Also in Solomon Islands.
HABITAT: Epiphyte in montane forest; 700 m.
COLLECTIONS: *Graeffe* 1242 (HBG); *Whistler* 1183 (BISH, HAW), 3867 (BISH, HAW, K).

D. macrophyllum *A.Rich.* in Sert. Astrol.: 22, t.9 (1834). Type: New Guinea, *Lesson* s.n. (holo. P!).
For full synonymy see Cribb in Kew Bull. 38: 247 (1983).

A medium-sized to large epiphyte. Pseudobulbs clavate, or subclavate, 15–50 cm tall, 1.4–3 cm in diameter in upper part, 3- or more-leaved towards apex. Leaves coriaceous, spreading or suberect, elliptic, obtuse, 15–31 × 3.3–9 cm. Inflorescences subterminal, subdensely many-flowered, 10–40 cm long; bracts lanceolate, acute, 1.5–2.5 cm long. Flowers setosely hairy on outer side of sepals, ovary and pedicel, white to creamy or greenish cream, spotted on sepals and petals with maroon, striped on side lobes of lip with purple; pedicel and ovary 2.5–4 cm long. Dorsal sepal oblong-ovate to oblong-lanceolate, acute, 2.1–2.6 × 0.7–1 cm. Lateral sepals obliquely triangular, acute, 2.3–2.6 × 1.2–1.4 cm; mentum obliquely conical, 0.8–1 cm long. Petals oblanceolate or oblong-oblanceolate, acute, 1.8–2.2 × 0.8–1.1 cm. Lip recurved, 3-lobed, 1–2 × 0.8–1.1 cm; side lobes quadrate, erect; midlobe transversely oblong, apiculate, conduplicate; callus 3-ridged in basal half of lip, white. Column 3 mm long; foot 1 cm long.

DISTRIBUTION: Widespread from Java east to New Guinea, S. Philippines, Micronesia, the Solomon Islands, Vanuatu, New Caledonia and Fiji. The

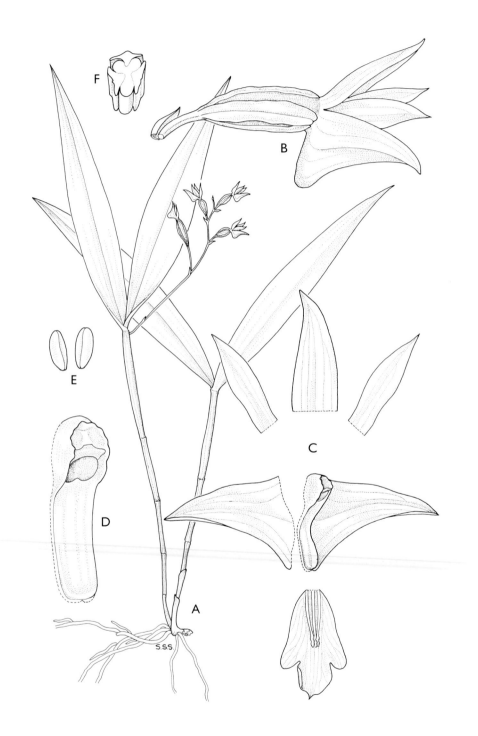

occurrence of this species in Samoa is debatable. The Graeffe collection may come from Fiji, where this showy orchid is common, rather than Samoa.

HABITAT: In montane forest; 600 – 700 m.

COLLECTIONS: *Graeffe* 1242 (HBG).

section CALYPROCHILUS

D. mohlianum *Rchb.f.* in Oesterr. Bot. Zeitschr. 20: 214 (1862). Type: Fiji, *Seemann* 578 (holo. K!).

D. neo-ebudanum Schltr. in Bull. Herb. Boiss. ser. 2, 6: 456 (1906). Type: Vanuatu, *Morrison* s.n. (holo. B†).

D. vitellinum Kraenzl. in Engl., Pflanzenr. Orch. Mon. Dendr. 1: 124, fig.7 (1910). Type: Vanuatu, *MacDonald* s.n. (holo. B†).

An epiphyte with clustered, erect to spreading stems. Stems many-noded, terete, 25–45 cm long, 2–6 mm in diameter, leafy in apical part, becoming ribbed with age. Leaves distichous, lanceolate to elliptic-lanceolate, subacute to acute, 5–11 × 1–2.4 cm, articulated to sheathing leaf bases, 1.3–3.5 cm long. Inflorescences lateral at upper nodes on leafless stems, short, racemose, densely 3–8-flowered, 3–4.5 cm long; peduncle 0.3–2 cm long; bracts ovate, chaffy, 5–13 mm long. Flowers orange-red to red, sometimes with a purple lip; pedicel and ovary 1.3–2.5 cm long. Dorsal sepal oblong-obovate, subacute to obtuse, 8–10 × 3–4 mm. Lateral sepals oblique, ovate, subacute, 15–23 × 3.5–5 mm; mentum retrorse, narrowly conical, 10–16 mm long. Petals oblanceolate to oblong-oblanceolate, obtuse, 8–10 × 3.5–4 mm. Lip clawed, entire, obovate, upcurved and crimped at the broad apex, 14–16 × 6–8 mm, serrate-fimbriate on apical margin. Column 3–4 mm long; foot 10–16 mm long.

DISTRIBUTION: Savai'i and 'Upolu. Also in the Solomon Islands, Vanuatu and Fiji.

HABITAT: Occasional epiphyte found in montane and cloud forests; 1000–1780 m.

COLLECTIONS: *Christophersen* 829 (BISH, K); *Reinecke* 437 (B†); *Vaupel* 180 (K), 578 (K); *Whistler* 2490 (BISH, HAW, K), 2543 (HAW), 2667 (HAW), 3957 (BISH, HAW), 8876 (HAW).

Fig. 16. *Dendrobium whistleri*. **A**, habit × ⅔; **B**, flower, side view × 3⅓; **C**, flower, dissected view with lip flattened × 3⅓; **D**, column × 7; **E** pollinia × 7; **F**, ovary transverse section × 3⅓. All drawn from the type collection by Susanna Stuart-Smith.

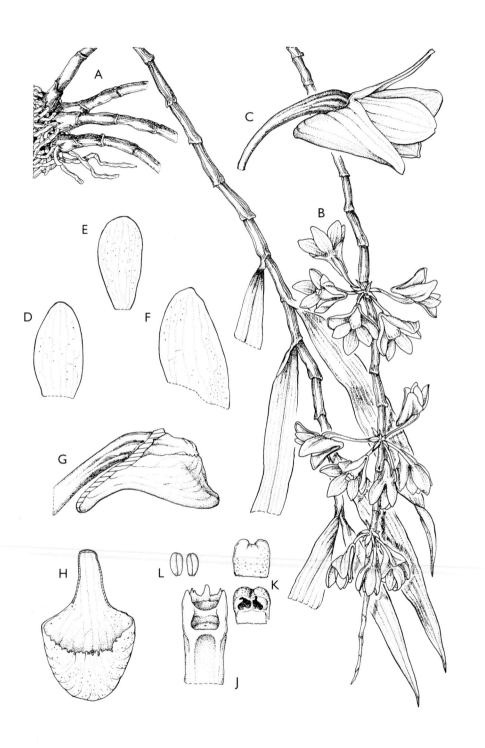

section SPATULATA

D. samoense *Cribb* in Kew Bull. 37: 584 (1983). Type: Samoa, Savai'i, *Vaupel* 282 (holo. K!, iso. BISH!, MO!).

D. tokai sensu Schltr. in Repert. Spec. Nov. Regni Veg. 9: 103 (1910); Setchell, American Samoa 1: 103 (1924); Christophersen in Bernice P. Bishop Mus. Bull. 128: 67 (1935) non Rchb.f.

D. sylvanum sensu Kraenzl. in Bot. Jahrb. Syst. 25: 605 (1898); H.Fleischm. & Rech. in Denkschr. Kaiserl. Akad. Wiss., Math.-Naturwiss. Kl. 85: 259 (1910), non Rchb.f.

An epiphyte up to 50 cm or more tall. Stems, cane-like, clustered, leafy. Leaves distichous, coriaceous, elliptic or ovate-elliptic, obtuse or rounded at apex, 5.5–9 × 3–3.5 cm; sheaths tubular, 1–3 cm long. Inflorescences erect, laxly 10–20-flowered, 20–45 cm long; bracts ovate-triangular, acute, 4–6 mm long. Flowers yellow or cream-coloured , not opening widely, probably self-pollinating; pedicel and ovary 2.5–3 cm long. Dorsal sepal narrowly oblong-lanceolate, acute, 1.8–2.5 × 3–4 mm. Lateral sepals obliquely lanceolate, acute, 1.8–2.2 × 0.5–0.6 cm; mentum obliquely conical, 7–8 mm long. Petals narrowly oblong-oblanceolate, acute, 1.8–2.4 × 0.2–0.35 mm, untwisted. Lip 3-lobed in apical half, 1.8–2 × 0.5–0.8 cm; side lobes narrowly oblong, erect, somewhat erose on front margins; midlobe ovate, acute, 9 × 5 mm, with an obscurely erose margin; callus of 3 longitudinal lamellae from base on basal half of midlobe, each raised into a flap at apex. Column 3–4 mm long; foot 6–7 mm long.

DISTRIBUTION: Savai'i, Tutuila and 'Upolu. Endemic to Samoa.

HABITAT: Occasional epiphyte from the littoral, lowland and montane forests; sea level to 600 m.

COLLECTIONS: *Bristol* 2096 (BISH); *Christophersen* 1923 (BISH, K); *Cox* 90 (BISH), 845 (BISH); *Garber* 619 (BISH); *Graeffe* 1263 in part (HBG); *Powell* 356 (K); *Rechinger* 91 (W); *Reinecke* 231 (B†); *Setchell* 354 (UC); *Sledge* 1701 (K); *Vaupel* 282 (B, BISH, K, MO); *Whistler* 132 (HAW), 203 (BISH, HAW), 487 (HAW), 812 (HAW), 1200 (HAW), 2795 (HAW, K), 4541 (HAW, K), 5174 (HAW), 5346 (HAW), 6878 (HAW), 8120 (HAW), 8667 (HAW), 8813 (HAW); *Whitmee* 42 (K); *Yuncker* 9066 (BISH), 9304 (BISH).

section RHOPALANTHE

D. goldfinchii *F.Muell.* in South Sci. Rec. 3: 4 (1883). Type: Solomon Islands, *Goldfinch* s.n. (holo. MEL).

D. crumenatum sensu H.G.Jones in Philipp. J. Sci. 104: 90 (1975), non Sw.

Aporum goldfinchii (F.Muell.) Brieger in Schltr., Die Orchideen ed.3, 1(11–12): 673 (1981).

Fig. 17. *Dendrobium mohlianum*. **A**, habit × ⅔; **B**, inflorescence × ⅔; **C**, flower × 2; **D**, dorsal sepal × 2; **E**, petal × 2; **F**, lateral sepal × 2; **G**, lip, column and ovary × 2; **H**, lip × 2; **J**, column from beneath × 3; **K**, anther cap, two views × 4; **L**, pollinia × 4. All drawn from *Wickison* 70 by Sue Wickison.

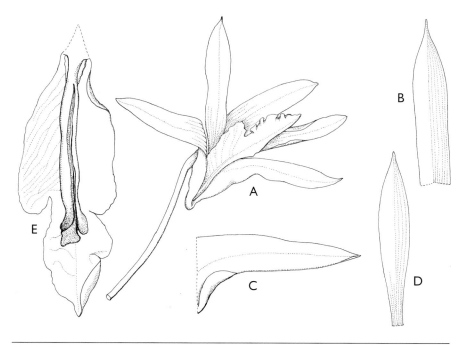

Fig. 18. *Dendrobium samoense.* **A**, flower × 1.5; **B**, dorsal sepal × 1.5; **C**, lateral sepal × 1.5; **D**, petal × 1.5; **E**, lip × 3. All drawn from *Whitmee* s.n. by Mair Swann.

An epiphyte with erect to pendent, clustered pseudobulbous stems. Stems swollen in basal part, tapering above, up to 50 cm long, yellow-green often tinged red or purple. Leaves distichous, bilaterally flattened, linear-lanceolate, acute, 3–7 × 4–8 mm, articulated to sheathing base. Inflorescences slender, laxly many-flowered, up to 25 cm long; bracts 4–5 mm long. Flowers appearing in succession, short-lived, somewhat translucent, pale creamy yellow to white with a yellow or green spot on the lip; pedicel and ovary 9–12 mm long. Dorsal sepal ovate, acute, 5–10 mm long, 3–4 mm wide. Lateral sepals very obliquely ovate, acute, 7–13 mm long and wide; mentum prominent, saccate, 6–14 mm long. Petals linear, acute, 4–8 × 1–2 mm. Lip recurved in upper half, 3-lobed, 7–18 × 6–12 mm; side lobes erect, obliquely oblong, rounded in front; midlobe transversely oblong-elliptic, deeply retuse; callus of 3 raised ridges from base to middle of midlobe, terminating in papillae. Column 3–4 mm long; foot 7–17 mm long.

Fig. 19. *Dendrobium goldfinchii.* **A**, habit × ⅔; **B**, flower × 3; **C**, dorsal sepal × 6; **D**, petal × 6; **E**, lateral sepal × 6; **F, G**, lip, two views × 6; **H**, column from beneath × 6; **J**, anther cap, two views × 8; **K**, pollinia × 8; **L**, flower × 3; **M**, dorsal sepal × 4; **N**, petal × 4; **O**, lateral sepal × 4; **P, Q**, lip, two views × 4; **R**, column from beneath × 4; **S**, anther cap, two views × 6; **T**, pollinia × 6. **A–K** drawn from *Wickison* 10; **L–T** from *Wickison* 11. All drawn by Sue Wickison.

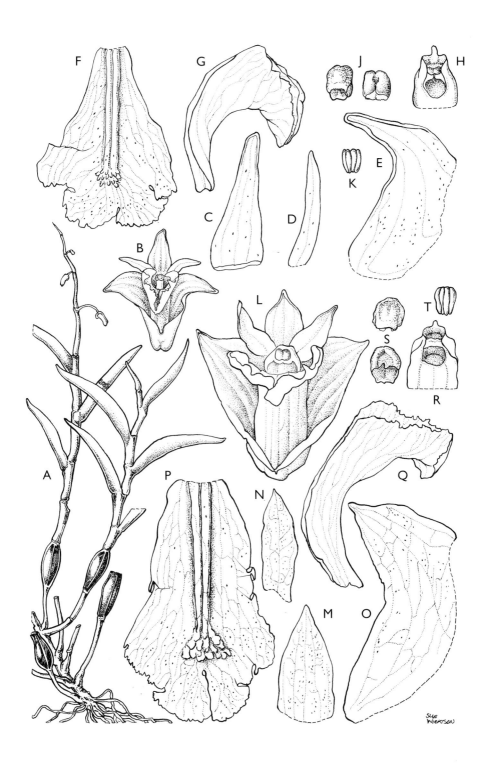

DISTRIBUTION: Apolima, Tutuila and 'Upolu. Also found in New Guinea, Bougainville, the Solomon Islands, Vanuatu and the Santa Cruz Islands, probably a recent arrival in Samoa.

HABITAT: Occasional epiphyte found in coastal and lowland forests; sea level to 300 m.

COLLECTIONS: *Whistler* 2895 (BISH, HAW, K), 3346 (HAW), 3622 (BISH, HAW, K), 4180 (BISH, HAW), 8060 (HAW), 8668 (BISH, HAW), 9103 (HAW) .

D. scirpoides *Schltr.* in Bot. Jahrb. Syst. 9: 103 (1911). Type: Samoa, 'Upolu, *Betche* s.n. (holo. B†).
Aporum scirpoides (Schltr.)S.Rauschert in Repert. Spec. Nov. Regni Veg.
94(7–8): 442 (1983).

An epiphyte, 15–20 cm tall; rhizome abbreviated, many-rooted. Stem with the second node above the base swollen, fusiform and sulcate, slender above, laxly 5–8-leaved. Leaves erect-spreading, somewhat subfalcate, subulate, 3.5–8 × 0.1–0.15 cm, articulated to sheathing bases somewhat shorter than the nodes. Inflorescence with the flowers produced singly but in the manner of *D. crumenatum* Sw. Flowers similar in size to those of *D. clavipes* Hook.f.; pedicel and ovary glabrous, 7 mm long. Sepals ovate-lanceolate, acuminate, 5 mm long. Lateral sepals oblique, forming a conical obtuse mentum, 5 mm long. Petals obliquely linear, acute, slightly dilated towards the base, 5 mm long. Lip cuneate at the base, entire, elliptic, shortly acuminate, glabrous, 9 × 4 mm, front margin undulate; callus of 2 parallel lamellae from the base to above the middle. Column short, tridentate at the apex; foot 5 mm long.

DISTRIBUTION: 'Upolu only. Endemic.
HABITAT: Unknown
COLLECTION: *Betche* s.n. (B†).

NOTE: Unfortunately, the type collection of *D. scirpoides* was destroyed in the Berlin Herbarium and we have not seen any other specimen that has been collected subsequently that agrees with it. It seems likely from his description of the habit and flowers that it fits within sect. *Rhopalanthe* (formerly sect. *Crumenatum*) but *D. goldfinchii*, the only other Samoan species in the section, has bilaterally flattened, rather than subulate leaves, and a 3-lobed lip with a trilamellate callus.

section GRASTIDIUM

D. biflorum *(G.Forst.)Sw.* in Nova Acta Reg. Soc. Sci. Upsal. 6: 84 (1799). Type: Tahiti, *J. & G. Forster* s.n. (holo. BM!, iso. K!).
Epidendrum biflorum G.Forst., Fl. Ins. Austr. Prodr.: 60 (1786).
Dendrobium vaupelianum sensu Yuncker in Bernice P. Bishop Mus. Bull. 184: 32 (1945), non Kraenzl.

Erect or ascending epiphytic plants . Stems slender, reed-like, many-noded, 35–160 cm long, 2–4 mm in diameter, leafy in upper part. Leaves distichous, linear-lanceolate, obliquely acute, 7–16 × 4–7 mm, articulated at base to tubular sheaths, 1–1.4 cm long. Inflorescences 2-flowered, borne at nodes opposite leaves; peduncle 3–7 mm long; bracts narrowly triangular, 1–1.5 mm long. Flowers ephemeral, pale yellow; pedicel and ovary 7–12 mm long. Dorsal sepal linear-lanceolate, filiform at apex, 28–32 × 1.5–2 mm. Lateral sepals obliquely linear-lanceolate, filiform at apex, 30–34 × 2–3 mm; mentum narrowly conical, 4–6.5 mm long. Petals linear-attenuate, filiform, 28–32 × 1–1.5 mm. Lip 3-lobed, recurved, 6–8 mm long, 3.5–4.5 mm wide; side lobes obliquely narrowly triangular, acute, erose on margins; midlobe narrowly triangular, acute, 3–4 × 2–2.5 mm, irregularly fimbriate on lower margins; callus a longitudinal raised keel from base of lip merging into 3 rows of papillae on basal part of midlobe. Column 1.5–2 mm long; foot 4.5–6 mm long.

DISTRIBUTION: Ofu, Savai'i, Ta'u, Tutuila and 'Upolu. Also in Bougainville, the Solomon Islands, the Santa Cruz Islands, Vanuatu, Fiji and Tahiti.

HABITAT: Common epiphyte in lowland and lower montane forest; sea level to 950 m.

COLLECTIONS: *Christophersen* 442 (BISH), 1880(BISH, K), 3008 (BISH); *Garber* 575 (BISH), 935 (BISH); *Kennedy* 3971 (BISH); *McKee* 3013 (BISH, K); *Powell* s.n. (K); *Rechinger* 699 (W), 1322 (W), 1803 (W); *Reinecke* 235 (B†); *Vaupel* 545 (B†); *Whistler* 284 (BISH, HAW), 485 (BISH, HAW), 1335 (BISH, HAW), 2791 (BISH, HAW, K), 2696 (BISH, HAW), 6998 (HAW), 7995 (HAW), 8123 (HAW), 8300 (HAW), 8781 (HAW), 9051 (HAW), 9520 (HAW), 9644a (HAW); *Yuncker* 9067 (BISH).

D. dactylodes *Rchb.f.* in J. Bot. 15: 132 (1877). Type: Samoa, *Whitmee* 46 (holo. W!, iso. K!, photo.of holo. AMES!).

D. involutum sensu Kraenzl. in Bot. Jahrb. Syst. 25: 603 (1898), non Lindl.

D. vaupelianum Kraenzl. in Notizbl. Konigl. Bot. Gart. Berlin 5: 109 (1909). Type: Samoa, *Vaupel* 286 (holo. B†, iso. AMES!, K!, W!).

D. whitmeei Kraenzl. in Engler, Pflanzenr. Orch. Mon. Dendr. 1: 189 (1910). Types: *Whitmee* s.n. (syn. B†), *Betche* 227 (syn. B†)., **synon. nov. e descr.**

D. everardii Rolfe in Bull. Misc. Inform., Kew 1921: 55 (1921). Types: Viti Levu, *Im Thurn* 316 (syn. K!), 326 (syn. K!).

D. cheesmanae Guillaumin in Bull. Soc. Bot. France 103: 280 (1956). Type: New Hebrides, *Cheeseman* A22 (holo. BM!)

An epiphyte with clustered slender stems. Stems 25–140 cm long, leafy, many-noded. Leaves distichous, lanceolate, obliquely obtuse or subacute at apex, 3.5–8 × 1–2.5 cm, articulated to sheathing leaf bases, 1–3 cm long. Inflorescences lateral, 2-flowered, borne at nodes opposite leaves, 3–6 mm long; bracts small, scale-like. Flowers white to pale creamy white, ephemeral; pedicel and ovary 7–13 mm long. Dorsal sepal linear-lanceolate, acute, 15–24

× 3–3.5 mm. Lateral sepals falcate, lanceolate, attenuate, acute, 13–28 × 6–10 mm; mentum inflexed, conical, 5–7 mm long. Petals linear-lanceolate, acute, 13–24 × 1–2 mm. Lip 3-lobed, recurved, 7–14 × 5–8 mm; side lobes erect, obliquely ovate, subacute to obtuse; midlobe ovate, attenuate, acuminate, sparsely papillate at apex of callus, the margins irregularly crenate-dentate below; callus with a prominently raised keel, verruculose in basal part, crenate in apical part. Column 3–3.5 mm long; foot 4.5–6 mm long.

DISTRIBUTION: Apolima, 'Aunu'u, Ofu, Olosega, Savai'i, Ta'u, Tutuila and 'Upolu. Also in Vanuatu, Fiji, Tonga and Cook Islands.

HABITAT: Common epiphyte in coastal to cloud forests; sea level to 800 m.

COLLECTIONS: *Christophersen* 441 (BISH, K), 1252 (BISH), 2256 (BISH), 3233 (BISH); *Dumas* 22 (P); *Garber* 557 (BISH); *Graeffe* 1265 (HAW, HBG), 1266 (HAW, HBG); *Hagner* 4007 (S); *Hellqvist* 3052 (S); *Powell* 251 (K), s.n. (K); *Rechinger* 494 (K, W), 522 (W), 882 (W), 1010 (W), 5266 (W); *Reinecke* 218 (B†), 232 (B†); *Setchell* 205 (UC), 205a (UC), 247 (UC); *Vaupel* 286 (AMES, K, W); *Whistler* 154 (BISH), 205 (HAW), 481 (HAW), 971 (HAW), 2695 (HAW), 2751 (HAW), 2831 (HAW), 2919 (BISH, HAW, K), 3081 (HAW, K), 3201 (HAW), 3803 (HAW, K), 3867 (HAW), 4382 (HAW), 4445 (HAW, K), 5170 (HAW), 5345 (HAW, K), 5734 (HAW), 6993 (HAW), 7001 (HAW), 7022 (HAW), 7055 (HAW), 7608 (HAW), 7983 (HAW), 8058 (HAW), 8119 (HAW), 9232 (HAW), 9384 (HAW), 9388 (HAW), 9544 (HAW); *Whitmee* 46 (K, W); *Yuncker* 9441 (BISH), 9442 (BISH).

NOTE: Very close to *D. involutum* Lindl. from Tahiti and adjacent islands. Their relationship needs further exploration.

D. reineckei *Schltr.* in Repert. Spec. Nov. Regni Veg. 9: 102 (1910). Type: Samoa, *Reinecke* 234 (holo. B†, iso. BM! G! K!)

D. gemellum sensu Kraenzl. in Bot. Jahrb. Syst. 25: 604 (1898), non Lindl.

D. samoanum Schltr. in Repert. Spec. Nov. Regni Veg. 9: 102 (1910), nomen nud.

An epiphyte with slender reed-like stems up to 100 cm long, 2–4 mm in diameter, many-noded, leafy in apical part. Leaves distichous, linear-lanceolate, obliquely acute, 8–15 × 4–7 mm, articulated to tubular, weakly striate leaf bases, 1.5–3 cm long. Inflorescences lateral, emerging opposite the leaves at nodes, 2-flowered; peduncle 1–1.5 cm long; bracts triangular, 1 mm long. Flowers membranous, white spotted with pink or red, ephemeral; pedicel and ovary 6–9 mm long. Dorsal sepal lanceolate, filiform above, 30–40 × 1.5–2 mm. Lateral sepals obliquely lanceolate, filiform above, 33–40 × 2.5–3 mm; mentum obliquely conical, 4–5 mm long. Petals lanceolate, filiform above, 30–40 × 1.5–2 mm. Lip 3-lobed, 9–12 × 4–5 mm; side lobes erect, oblong-ovate; midlobe triangular, acute, with fimbriate side margins; callus weakly tricarinate at base, the central one extending onto base of midlobe. Column 1–1.5 mm long; foot 4–5 mm long.

DISTRIBUTION: Savai'i and 'Upolu. Also in Fiji.

HABITAT: Epiphyte in montane and cloud forest; 600–1550 m.

COLLECTIONS: *Reinecke* 234 (BM, G, K); *Whistler* 155 (HAW), 2607 (BISH, HAW, K), 3958 (HAW), 5173 (HAW), 5706 (HAW), 7021 (HAW), 9401 (HAW), 9598 (HAW).

D. sladei *J.J.Wood & Cribb* in Orchid Rev. 90: 14, fig.7 (1982). Type: Vanuatu, Efate, *Im Thurn* 330 (holo. K!).

A large epiphyte with clustered erect to pendent stems. Stems reed-like, hard, 50–122 cm long, leafy in upper half. Leaves distichous, narrowly ovate to lanceolate, unequally obtusely bilobed at apex, 5–11 × 1.5–2.8 cm, articulated to sheathing, 1.5–4 cm long leaf bases. Inflorescences lateral, opposite leaves from nodes, 2-flowered; peduncle up to 6 mm long; bracts scale-like. Flowers ephemeral, yellow to creamy yellow, the lip with reddish brown markings on side lobes; pedicel and ovary 10–12 mm long. Dorsal sepal lanceolate, acute, 16–25 × 2–4 mm. Lateral sepals obliquely lanceolate-falcate, acute, 17–30 × 4–8.5 mm; mentum incurved-conical, 5–6 mm long. Petals ligulate-lanceolate, acute, 16–25 × 2–4 mm. Lip recurved, 3-lobed, 9–12.5 × 5–7 mm; side lobes falcate-ovate, acute, minutely erose on margins; midlobe ovate, acute, 6 × 5 mm, sometimes sparsely papillose, the margins erose-undulate; callus linear, raised, from base of lip to middle of midlobe, verruculose at base, carinate above, irregularly crenate on upper margin. Column 4 mm long; foot 5–6 mm long.

DISTRIBUTION: Savai'i, Ta'u, Tutuila and 'Upolu. Also in Vanuatu and Fiji.

HABITAT: Occasional in montane forest; 360–600 m.

COLLECTIONS: *Bryan* 960 (BISH); *Christophersen* 106 (BISH), 145 (BISH), 1190 (BISH), 1881 (BISH), 3007 (BISH), 3166 (BISH), 3264 (BISH); *Diefenderfer* 21 (BISH); *Garber* 571 (BISH), 655 (BISH); *Rechinger* 137 (W); *Whistler* 71 (HAW), 441 (HAW), 583 (HAW), 3958 (BISH, HAW), 5167 (HAW), 5168 (HAW), 5171 (HAW), 5172 (HAW), 7054 (HAW), 7998 (HAW), 8136 (HAW), 8199 (HAW), 8497 (HAW), 9088 (HAW), 9385 (HAW); *Yuncker* 9067 (BISH), 9160 (BISH), 9385 (K), 9391 (BISH).

D. lepidochilum *Kraenzl.* in Engl., Pflanzenr. Orch. Mon. Dendr. 1: 187 (1910). Type: Samoa, 'Upolu, *Reinecke* 233 (holo. B†, iso. K!).

D. sp. 1 sensu Kraenzl. in Bot. Jahrb. Syst. 25: 603 (1898).

Dendrobium sp. 2; Gerlach in Gartenbauwissenschaft 57 (5): 220 (1992).

A large epiphyte with clustered pendent leafy stems up to 40 cm or more long. Leaves coriaceous, oblong-ovate, rounded at unequally bilobed apex, 6–12.5 × 1.5–2 cm. Inflorescences 2-flowered, very short; bracts obscure. Flowers opening widely, yellow speckled all over inside of sepals and petals with red, the lip side lobes red-striped; pedicel and ovary 7 mm long. Dorsal

sepal linear-lanceolate, subacute, 20–22 × 2–3 mm. Lateral sepals spreading-incurved, falcate, lanceolate, subacute, 22–23 × 6–7 mm. Petals linear, acute, 20–22 × 2–3 mm. Lip 3-lobed, 9–11 × 6–7 mm; side lobes erect, obscurely elliptic, acute; midlobe ovate, acute, scarcely papillate; callus a central longitudinal ridge. Column short, 1.5 mm long; foot 4–5 mm long.

DISTRIBUTION: Savai'i, Tutuila and 'Upolu. Possibly endemic.

HABITAT: Epiphyte in swamp and montane forest; 200–300 m.

COLLECTIONS: *Christophersen* 1256 (BISH), *Cox* 329 (BISH); *Rechinger* 1890 (W); *Reinecke* 233 (K); *Whistler* 2694 (HAW, K), 7000 (HAW), 8136 (HAW), 9038 (HAW), 9542 (HAW).

NOTE: The relationship of this species to *D. sladei* needs further investigation. Some of the sterile or badly preserved specimens attributed to *D. sladei* may better be placed here. The flowers of this orchid are illustrated with a fine colour photograph in Fig. 5 of Gerlach (1992a).

36. **DIPLOCAULOBIUM**

Kraenzl. in Engl., Pflanzenr. Orch. Mon. Dendr. 1: 331 (1910).

Small to large epiphytic herbs with short to elongate creeping rhizomes and erect dimorphic flask-shaped pseudobulbs, unifoliate at apex. Leaves somewhat coriaceous, linear to oblong, articulated at base. Inflorescences apical, fasciculate, 1–several-flowered. Flowers resupinate, small to large, ephemeral, often lasting less than a day, turning reddish with senescence. Sepals free, often drawn out into a filamentous or acuminate apex; laterals oblique, forming a more or less conspicuous mentum with the column-foot. Petals free, narrower than the sepals. Lip entire to 3-lobed, articulated to column-foot, with a ridged callus and frequently pubescent towards apex. Column short, with a prominent foot; pollinia 4, waxy, lacking an appendage.

A genus of about 70 species centred on New Guinea but extending westward to Malaya and eastward to Micronesia, New Caledonia, Fiji and Samoa. A single species has been reported from Samoa.

D. fililobum *(F. Muell.) Kraenzl.* in Engl., Pflanzenr. Orch. Mon. Dendr. 1: 334 (1910). Type: Samoa, *Betche* s.n. (holo. MEL!).

Dendrobium fililobum F.Muell. in South Sci. Rec.: 1 (1882).

Dendrobium tipuliferum sensu Yuncker in Bernice P. Bishop Mus. Bull. 184: 32 (1945) non Rchb.f.

Fig. 20. *Dendrobium lepidochilum.* **A,** habit × ½; **B,** flower × 1; **C,** flower with proximal lateral sepal, petal and lip removed × 3; **D,** dorsal sepal, petal and lateral sepal × 3; **E, F,** lip with detail of base × 6; **G,** column × 6; **H,** anther cap, two views × 10; **J,** pollinia × 10. All drawn from *Whistler* 7000 by Susanna Stuart-Smith.

An epiphyte with elongate pseudobulbs, 12–20 cm long, 3–4 mm wide, drying bright yellow. Leaf erect, linear, subacute to acute, 7.5–12.5 × 0.7–0.9 cm, dull green. Inflorescence one-flowered; bract linear-conduplicate, acute, 2–2.5 cm long. Flowers creamy white with purple on side lobes of lip, lasting a day; pedicel elongate, 5–6 cm long. Dorsal sepal ovate at base, with a long apical filament, 35–40 × 2 mm. Lateral sepals obliquely triangular at base, filamentous above, 38–42 × 4–5 mm; mentum obliquely conical, 4–5 mm long. Petals linear, filamentous, 3.3–4 cm long. Lip clawed at base, 3-lobed in basal half, 12–14 × 4–5 mm; side lobes falcate, triangular, subacute; midlobe longly clawed with a triangular apical lamina; callus of three keels, the outer two running from base to middle of lip, the central one short, in middle of lip only. Column 3 mm long.

DISTRIBUTION: Olosega, Ta'u, Tutuila, Savai'i and 'Upolu. Endemic to Samoa.

HABITAT: Epiphyte found in montane and cloud-forests; 400 – 1550 m.

COLLECTIONS: *Betche* s.n. (MEL); *Christophersen* 1203 (BISH), 1213 (BISH), 2276 (BISH), 3486 (BISH); *Graeffe* 1246 (HAW); *Kuruc* 10 (HAW); *Vaupel* 578 (B†); *Whistler* 484 (HAW), 710 (HAW, K), 1182 (HAW), 2024 (HAW), 2608 (HAW), 3095 (BISH, HAW), 3161 (HAW), 5155 (HAW), 6931 (BISH, HAW) , 7023 (HAW), 8002 (HAW), 8788 (HAW), 9196 (HAW), 9523 (HAW), 9650 (HAW); *Yuncker* 9264 (BISH).

37. **FLICKINGERIA**

A.D.Hawkes in Orchid Weekly 2: 451 (1961).

Small to large epiphytic herbs with creeping rhizomes often giving rise to superposed pseudobulbous stems. Stems many-noded, swollen distally, cylindric and slender in basal half, unifoliate. Leaf coriaceous, oblong to elliptic, articulated at base but lacking a basal sheath. Inflorescences 1- several, terminal or pseudoterminal, one-flowered, fasciculate. Flowers small to medium-sized, ephemeral, lasting usually less than a day. Dorsal sepal ovate to lanceolate, free. Lateral sepals obliquely triangular-ovate, forming a distinct mentum with the column-foot. Petals free, narrower than dorsal sepal. Lip articulated to column-foot, complex, 3-lobed; side lobes erect, more or less embracing column; midlobe more or less distinctly clawed, apical blade transverse, entire to lacerate-fimbriate. Column short, with a distinct foot; pollinia 4, waxy, lacking an appendage.

A genus of about 70 species in tropical Asia, South East Asia, the Malay Archipelago, New Guinea, N.E. Australia, the Pacific Islands from

---

Fig. 21. *Flickingeria comata*. **A**, habit, × ⅔; **B**, flower × 1.5; **C**, flower with proximal lateral sepal and petal removed × 3; **D**, dorsal sepal × 3; **E**, petal × 3; **F**, lateral sepal × 3; **G**, lip × 3; **H**, column × 4; **J**, anther cap, two views × 4. All drawn from *Wickison* 42 by Sue Wickison.

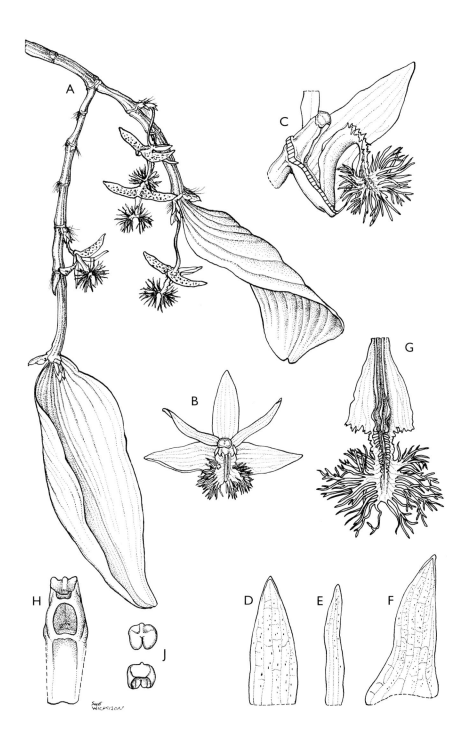

Micronesia south to New Caledonia and across to Samoa. A single species recorded from Samoa.

F. comata (*Blume*)*A.D.Hawkes* in Orchid Weekly 2: 453 (1961).Type: Java, *Blume* s.n. (holo. L, iso. P).
Desmotrichum comatum Blume, Bijdr. Fl. Ned. Ind.: 330 (1825).
Dendrobium comatum (Blume) Lindl., Gen. Spec. Orch. Pl.: 76 (1830).
D. thysanochilum Schltr. in K.Schum. & Lauterb., Nachtr. Fl. Deutsch. Schutzgeb. Sudsee: 152 (1905). Type: New Britain, *Schlechter* 13720 (holo. B†).
D. scopa sensu Setchell, American Samoa 1: 103 (1924), non Lindl.
Desmotrichum thysanochilum (Schltr.)Carr in Bull. Misc. Inform., Kew 1934: 380 (1934).
Ephemerantha comata (Blume) P.F.Hunt & Summerh. in Taxon 10: 106 (1961).

An erect to pendent much branched epiphyte. Stems superposed, clavate, up to 30 cm long, 1 cm in diameter, shiny, reddish brown turning bright yellow with age. Leaf elliptic, subacute, 11–15 × 3–5 cm. Inflorescences 1-flowered, terminal. Flowers lasting less than a day, sepals and petals pale yellow with red markings near the base, the lip brighter yellow. Dorsal sepal oblong-ovate, acute, 12–14 × 4 mm. Lateral sepals obliquely oblong-attenuate, acute, 12–14 × 5–6 mm; mentum 5–6 mm long. Petals linear-lanceolate, acute, 10–12 mm long, 1.5–2 mm wide. Lip prominently 3-lobed, 12–14 mm long; side lobes erect, small, subquadrate; midlobe oblong, 5.5–6.5 mm long, with elongate lateral filiform appendages near apex up to 10 mm long. Column short, 2.5–3 mm long; foot 4.5–5.5 mm long.

DISTRIBUTION: Savai'i, Ta'u, Tutuila and 'Upolu. S.E. Asia from Taiwan and Malaya, through Malesia to northern Australia, and eastwards to parts of Micronesia, the Solomon Islands, Vanuatu, New Caledonia and Fiji.
HABITAT: Epiphyte found in coastal, mangrove and montane forests; sea level to 500 m.
COLLECTIONS: *Bryan* 973 (BISH); *Christophersen* 1186 (BISH), 3526 (BISH); *Garber* 933 (BISH, K); *Powell* 363 (K); *Setchell* 335 (UC); *Vaupel* 283 (AMES, K); *Whistler* 482 (HAW), 2697 (HAW), 2829 (BISH, HAW, K), 2894 (BISH, HAW, K), 3162 (BISH, HAW, K), 7099 (HAW), 7794 (HAW), 8189 (HAW), 9041 (HAW), 9513 (HAW).

38. **PSEUDERIA**

Schltr. in Repert. Spec. Nov. Regni Veg. Beih. 1: 644(1912).

Large or medium-sized terrestrial or rarely epiphytic, scandent or climbing herbs with elongate slender, branching leafy stems. Leaves chartaceous-coriaceous, ovate to lanceolate, articulated to a sheathing tubular leaf base. Inflorescences short, lateral, racemose; peduncle base covered by imbricate sheaths. Flowers relatively small, non-resupinate, somewhat fleshy. Dorsal sepal free, ligulate. Lateral sepals somewhat falcate, shorter than the dorsal sepal. Petals free, erect, falcate, narrower than the sepals. Lip adnate to column-foot, immobile, somewhat recurved, entire, distally puberulent, with a longitudinal central callus. Column elongate, slender, slightly incurved; pollinia 4, waxy, unequal, lacking an appendage.

A small genus of about 18 species in New Guinea across to Micronesia, Fiji and Samoa. A single species reported from Samoa.

P. ramosa *L.O.Williams* in Bot. Mus. Leafl. 7: 140 (1939). Type: Samoa, *Christophersen* 3533 (holo. AMES!, iso. BISH!, K!).
P. sp. sensu Christophersen in Bernice P. Bishop Mus. Bull. 128: 67 (1935).

A large scrambling or climbing epiphyte with leafy stems up to 40 or more cm long. Leaves lanceolate, acuminate, $10–15 \times 1.4–2.5$ cm, articulated to sheathing leaf bases. Inflorescences many, emerging opposite the subtending leaves, 3–4 cm long, laxly to subdensely 3–4-flowered; bracts subimbricate, ovate, up to 10 mm long. Flowers creamy white to greenish, fleshy; pedicel and ovary 6–10 mm long. Dorsal sepal incurved, linear-oblanceolate, obtuse, 11–12 $\times$ 1.5–2 mm. Lateral sepals falcate, lanceolate, obtuse, $12–14 \times 3–4$ mm. Petals falcate, lanceolate, obtuse, $10–11 \times 1.5$ mm. Lip entire, elliptic-rhombic, subacute, $8–9 \times 4–5$ mm, glandular-farinose in apical half; callus basal, triangular, sulcate. Column 5–6 mm long.

DISTRIBUTION: Savai'i, Ta'u and Tutuila. Also found on Futuna.

HABITAT: Occasional epiphyte found in montane forest and scrub forest; 270–700 m.

COLLECTIONS: *Christophersen* 1190a (BISH, P), 3533 (AMES, BISH, K); *Garber* 717 (BISH); *Whistler* 3183 (HAW), 3345 (BISH, HAW, K), 5151 (HAW), 7988 (HAW); *Yuncker* 9251 (BISH).

39. **BULBOPHYLLUM**

Thouars, Hist. Orchid.: t.3 sub u. (1822).

Small to large epiphytic, lithophytic or very rarely terrestrial herbs with short to long creeping rhizomes. Stems pseudobulbous or rarely almost absent, unifoliate (or sometimes bifoliate outside our region) at apex. Leaf coriaceous to chartaceous, rarely deciduous. Inflorescences solitary or fasciculate, lateral from base of pseudobulb or from rhizome, 1–many-flowered, racemose, subumbellate or subcapitate. Flowers small to relatively large and showy, sessile to long-pedicellate. Dorsal sepal free or rarely adnate to lateral sepals. Lateral sepals adnate to column-foot at base to form a more or less prominent mentum, free or partly connate above. Petals free, narrower than sepals. Lip articulate to column-foot, mobile, rarely not, entire to 3-lobed, often very fleshy and recurved, often ciliate or pubescent, usually with a callus. Column short with a more or less prominent foot; pollinia 2 or 4, waxy, usually coherent in two pairs.

A large genus of possibly 1500 species found throughout the tropics and subtropics but best represented in S.E. Asia and Malesia. Eleven species have been reported from Samoa.

1. Inflorescence apparently umbellate; lateral sepals much longer than the dorsal sepal, connate in upper part · · · · · · · · · · · · · · **B. longiflorum**
   Inflorescence racemose or simple; lateral sepals not as above · · · · · · · **2**
2. Inflorescence one-flowered · **3**
   Inflorescence racemose, two- or more-flowered · · · · · · · · · · · · · · · **7**
3. Inflorescence sessile or subsessile · **4**
   Inflorescence elongate, pedunculate · **5**
4. Leaves 3.5–5 × 0.5–1 cm; inflorescences 0.5–1 cm long; flowers purple; sepals 5 mm long; lip 2.5–3 mm long · · · · · · · · · · · · · · · · · **B. betchei**
   Leaves 2–4 × 1–1.5 cm; inflorescence 2 mm long; flowers pale translucent yellow with a purple lip; sepals 2.5–3 mm long; lip 1.5 × 0.8 mm · **B. membranaceum**
5. Flowers large; sepals more than 25 mm long; petals more than 10 mm long, filamentous · **B. trachyanthum**
   Flowers smaller; sepals less than 15 mm long; petals less than 10 mm long · **6**
6. Lip pubescent; petals aristate; flowers yellow · · · · · · **B. distichobulbum**
   Lip glabrous; petals not aristate; flowers purplish · · · · · · ·**B. samoanum**
7. Inflorescence subglobose · **B. atrorubens**
   Inflorescence laxly 3- or more-flowered · · · · · · · · · · · · · · · · · · · **8**
8. Flowers small; sepals 6 mm long or less · · · · · · · · · · · · · · · · · · · **9**
   Flowers large; sepals 20 mm or more long · · · · · · · · · · · · · · · · · **10**
9. Inflorescence 9–14 cm tall, 15–20-flowered · · · · · · · · · · · · · **B. ebulbe**
   Inflorescence 2–5.5 cm tall, 3–6-flowered · · · · · · · · · · · · **B. savaiense**
10. Petals acute, 12–16 mm long; lip lacking horn-like appendage near base · **B. pachyanthum**
    Petals small, acuminate, c. 2 mm long; lip with two erect horn-like appendages near base · **B. longiscapum**

section APHANOBULBON

B. ebulbe *Schltr.* in K. Schum & Lauterb., Nachtr. Fl. Deutsch. Sudsee: 200 (1905). Type: Punam, *Schlechter* 14639 (holo. B †).

B. polypodioides Schltr. in Bot. Jahrb. Syst. 39: 86 (1906). Type: New Caledonia, *Schlechter* 15422 (holo. B †; iso. AMES!, K!, P!, Z!).

B. nigroscapum Ames, Orchidaceae 7: 86 (1922). Type: Samoa, *Setchell* 383 (holo. AMES!).

B. sp. 3 sensu Christophersen in Bernice P. Bishop Mus. Bull. 128: 68 (1935).

A creeping epiphyte with a long flexuose rhizome, 2–4 mm in diameter. Stems non-pseudobulbous, widely spaced along rhizome, 2–7 × 2–3 mm. Leaf oblong-ligulate, obtuse or slightly retuse at apex, 5–12.5 × 1–2.3 cm, narrowly petiolate. Inflorescence erect, racemose, 9–14 cm long; peduncle slender, terete, bearing 2 to 3 widely spaced sheaths; rachis 1.5–2 times length of peduncle, 15–20-flowered; bracts lanceolate, acuminate, 5–6 mm long. Flowers white, glabrous; pedicel and ovary 3–4 mm long. Dorsal sepal oblong-lanceolate, acute, 4–5 × 1–1.5 mm. Lateral sepals obliquely ovate, acute, 5–6 × 2.5–3 mm. Petals narrowly elliptic-lanceolate, subacute to obtuse, 2.5–3 × 0.8–1 mm, subentire or irregularly denticulate on margins. Lip ligulate, arcuate, subacute to obtuse, 2.5–4 × 0.9 mm, prominently channelled at base. Column very short, 1.5 mm long; foot slightly incurved, 1.2 mm long.

DISTRIBUTION: Savai'i, Tutuila and 'Upolu. Also found in the Solomon Islands, New Caledonia and Fiji.

HABITAT: Uncommon epiphyte found in lowland and montane forest; sea level to 950 m.

COLLECTIONS: *Christophersen* 418 (BISH), 1063 (BISH), 1181 (BISH); *Setchell* 383 (AMES, UC); *Whistler* 1181 (BISH), 2026 (K), 3244 (BISH, HAW, K), 3920 (BISH, HAW), 4166 (HAW), 8138 (HAW), 8503 (HAW), 8539 (HAW), 8726 (HAW), 8737 (HAW), 9005 (HAW), 9512 (HAW).

section DIALEIPANTHE

B. longiscapum *Rolfe* in Bull. Misc. Inform., Kew 1896: 45 (1896). Type: Fiji, cult. Kew, *Yeoward* s.n. (holo. K!).

B. praealtum Kraenzl. in Notizbl. Konigl. Bot. Gart. Berlin 5: 109 (1909). Type: Samoa, *Vaupel* 322 (holo. B!, iso. K!).

A large creeping epiphyte with an elongated rhizome 3–4.5 mm in diameter. Pseudobulbs widely spaced on rhizome, narrowly conical-ovoid, 2–4.5 × 0.8–1.8 cm. Leaf oblong-elliptic, broadly acute, 11.5–23 × 2–3.5 cm, petiolate. Inflorescence erect-arcuate, racemose, 25–79 cm long; peduncle wiry, terete; rachis somewhat fractiflex; bracts broadly ovate, acuminate, 8–12 mm long. Flowers produced sequentially, white or greenish yellow marked with dull purple or red at base of segments; pedicel and ovary 10–15 mm long. Dorsal sepal lanceolate, acuminate, 25–30 × 5–5.5 mm. Lateral sepals weakly spreading, obliquely lanceolate, attenuate, acute, 30–35 × 6.5–7.5 mm. Petals broadly ovate, acuminate, briefly setiform at apex, 2 × 1.5 mm. Lip very fleshy, porrect, oblong-lanceolate, narrowly obtuse, 26–32 × 8–9 mm; side margins revolute, crenate-undulate distally; callus of two prominently raised keels which coalesce in front. Column 5 mm long; foot 7–8 mm long.

DISTRIBUTION: Savai'i, Ta'u, Tutuila and 'Upolu. Also known from the Solomon Islands, Vanuatu, Fiji, Tonga, Niue and Wallis Island.

HABITAT: Occasional epiphyte in coastal to montane forest; sea level to 500 m.

COLLECTIONS: *Christophersen* 592 (BISH), 1037 (BISH, K); *Setchell* 274; *Vaupel* 322 (B, K); *Whistler* 343 (BISH, HAW), 1550 (HAW), 2827 (BISH, HAW, K), 3246 (HAW), 3887 HAW), 6795 (HAW), 7997 (HAW), 8059 (HAW), 8724 (HAW), 9003 (HAW); *Yuncker* 9076 (BISH).

B. pachyanthum *Schltr.* in Bot. Jahrb. Syst. 39: 85 (1906). Type: New Caledonia, *Schlechter* 15678 (holo. B†).

B. longiscapum sensu B.E.V.Parham in Trans. Proc. Fiji Soc. 2: pl.11 (1953); non Rolfe.

B. sp. 2 sensu Yuncker in Bernice P. Bishop Mus. Bull. 184: 33 (1945).

A large epiphyte with a clustered or weakly spreading habit; rhizome somewhat elongated, 4–6 mm in diameter. Pseudobulbs closely spaced, obliquely ovoid, 1.8–3.5 × 1–1.7 cm. Leaf oblong-elliptic, broadly acute, 8–22 × 3–5.2 cm, narrowly petiolate at base. Inflorescence erect, racemose, 22–40 cm long, laxly 2–4-flowered; peduncle slender, terete, bearing 3–4 widely spaced sheaths along its length; rachis short; bracts broadly ovate, acuminate, 10–12 mm long. Flowers developing sequentially, green or greenish yellow spotted with purple, glabrous; pedicel and ovary 1.5–2.5 cm long. Dorsal sepal narrowly ovate, acuminate, 2.2–2.8 × 1.2–1.4 cm, slightly thickened and carinate dorsally. Lateral sepals obliquely lanceolate, acute, 3.3–3.8 × 0.8–1.2 cm, dorsally carinate. Petals obliquely ovate, acute, 1.2–1.6 × 0.8–0.9 cm. Lip arcuate, fleshy, oblong-lanceolate, subacute, 1.2–1.4 × 0.5 cm, auriculate at base; calli 2, longitudinal, somewhat papillate. Column 1 cm long, with setose stelidia as long as the column; foot 0.8 cm long.

DISTRIBUTION: Savai'i, Ta'u, Tutuila and 'Upolu. Also from New Caledonia, Fiji and Tonga.

HABITAT: Uncommon epiphyte in montane scrub and forest; 300 to 1600 m.

COLLECTIONS: *Christophersen* 558 (BISH), 1044 (BISH), 1179 (BISH), 1212 (BISH); *Whistler* 3344 (HAW), 9614 (HAW) *Yuncker* 9250 (BISH).

section TRACHYANTHUM

B. trachyanthum *Kraenzl.* in Oesterr. Bot. Zeitschr. 44: 336 (1894). Type: New Ireland, *Micholitz* s.n. (holo. W!).

An epiphyte with clustered pseudobulbs on a short rhizome, c. 3 mm in diameter. Pseudobulbs 3–10 mm apart, narrowly conical to conical-ovoid, weakly 4-angled, 1.8–3 × 0.7–1.3 cm. Leaves oblong-lanceolate to oblong-

---

Fig. 22. *Bulbophyllum longiscapum*. **A**, habit × ⅔; **B**, flower × 1.5; **C**, dorsal sepal × 2; **D**, petal × 4; **E**, lateral sepal × 2; **F**, lip and column × 3; **G**, anther cap × 6; **H**, pollinia × 6. **A** drawn from *Wickison* 51; **B–H** from *Bregulla* 10. All drawn by Sue Wickison.

elliptic, acute or subacute, 7.5–13 × 1.2–2.2 cm, slenderly petiolate. Inflorescences erect, one-flowered, up to 20 cm long; peduncle 7–14 cm long; bract clasping, ovate, acuminate, much shorter than ovary. Flower large, the sepals green to greenish brown with numerous purple blotches, the petals yellow-green with purple tips, the lip greenish yellow; pedicel and ovary up to 4 cm long. Dorsal sepal lanceolate, linear-acuminate at apex, 28–35 × 5 mm. Lateral sepals falcate, lanceolate, acuminate, 25–30 × 4–4.5 mm. Petals ovate at base, linear-clavate at apex, 12–14 × 3–4 mm. Lip fleshy, arcuate, oblong-ligulate, obtuse, 5 × 3 mm, weakly sulcate at base. Column 3–4 mm long, with weakly quadrate stelidia; foot c. 3 mm long.

DISTRIBUTION: Savai'i. Also in New Guinea, the Solomon Islands and Fiji.
HABITAT: Rare in montane and cloud-forest; 400–1030 m.
COLLECTIONS: *Christophersen* 2188 (BISH), 3177 (BISH).
NOTE: Both the cited collections are sterile but have been placed here by Kores (1991). We are inclined to agree with his determination. Floral description based on Fijian material.

section FRUTICICOLA

B. betchei *F.Muell.* in South Sci. Rec. 1:173 (1881). Type: Samoa, *Betche* 261 (holo. MEL!).
B. finetianum Schltr. in Bot. Jahrb. Syst. 39: 83 (1906). Type: New Caledonia, *Schlechter* 15416 (holo. B†, iso. K!, P!, Z!).
B. atroviolaceum H.Fleischm. & Rech. in Denkschr. Kaiserl. Akad. Wiss., Math.-Naturwiss. Kl. 85: 261 (1910). Type: Samoa, *Rechinger* 1824 (holo. W!).
B. ponapense Schltr. in Bot. Jahrb. Syst. 56: 484 (1921). Type: Caroline Is., *Ledermann* 13447 (holo. B†).

Hanging epiphytic plants with elongated simple or branching rhizomes up to 18 cm long, 1–1.5 mm in diameter. Pseudobulbs closely spaced, ovoid, 4–7 × 1.5–3 mm. Leaves oblong, oblong-elliptic or oblong-obovate, acute or subacute, 2–4.6 × 0.4–0.9 cm, dark olive-green, shortly petiolate. Inflorescences short, one-flowered, 8–15 mm long; peduncle slender, terete, 3–7 mm long, bearing 2–3 sheaths; bract small, tubular-infundibuliform, c. 2 mm long. Flowers bright yellow with maroon stripes, the lip yellowish green; pedicel and ovary 4–6 mm long. Dorsal sepal lanceolate to elliptic-lanceolate, acuminate,

---

Fig. 23. *Bulbophyllum betchei*. **A**, habit × ⅔; **B**, flower × 6; **C**, dorsal sepal × 6; **D**, lateral sepal × 6; **E**, petal × 6; **F**, column and lip × 10; **G**, lip × 10; **H**, column × 14; **J**, anther cap × 14; **K**, pollinia × 14. **A** drawn from *Mackee* 19; **B–K** from *Sprunger* 206. *B. samoanum*. **L**, habit × ⅔; **M**, flower × 3; **N**, dorsal sepal × 4; **O**, lateral sepal × 4; **P**, petal × 4; **Q**, lip × 6; **R**, column and lip × 6; **S**, column from beneath × 6; **T**, anther cap, two views × 10; **U**, pollinia, three views × 10. **L** from *Whistler* 2787; **M–U** from *Sprunger* 121. All drawn by Sue Wickison.

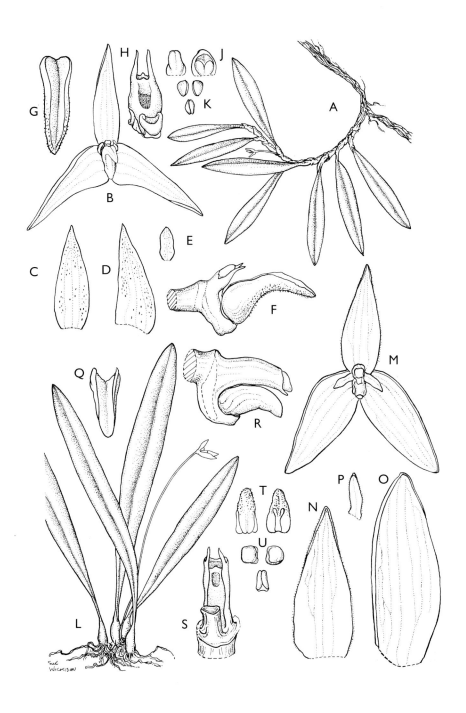

3.5–5 × 1–1.5 mm. Lateral sepals obliquely lanceolate, acute, 4–5.5 × 1.5–2 mm. Petals obliquely oblong, obtuse, 1.5 × 0.5–0.6 mm. Lip very fleshy, slightly arcuate, oblong-ligulate, obtuse, 2–2.6 × 1 mm, with 2 raised puberulent keels on upper surface. Column 1.5–2 mm long; foot 0.5 mm long.

DISTRIBUTION: Olosega, Savai'i, Ta'u and 'Upolu. Also in the Caroline Islands, Solomon Islands, Vanuatu and New Caledonia .

HABITAT: Occasional epiphyte in montane forest; 400 to 750 m.

COLLECTIONS: *Betche* 261 (MEL); *Christophersen* 37 (BISH), 174 (BISH), 2244a (BISH), 2245 (BISH, K); *Rechinger* 105 (W), 441 (W), 1617 (W), 1824 (W); *Vaupel* 537 (B†); *Whistler* 164 (HAW), 711 (HAW), 766 (HAW), 900 (BISH, HAW), 2046 (BISH, HAW), 3727 (BISH, HAW), 3817 (HAW, K), 5145 (HAW), 7085 (HAW), 8294 (HAW), 9397 (HAW).

section GLOBICEPS

B. atrorubens *Schltr.* in Bot. Jahrb. Syst. 39: 82 (1906). Type: New Caledonia, *Schlechter* 15495 (holo. B†, iso. K!, P!).

A small epiphyte with a short creeping rhizome. Pseudobulbs clustered, narrowly ovoid, 4–6 × 2–4 mm. Leaves erect, obovate, obtuse and minutely tridenticulate at apex, 4–20 × 1.5–2.3 cm, yellow-green; petiole c. 2 cm long. Inflorescence erect, clavate, densely many-flowered; peduncle slender, 10–20 cm long; rachis short; bracts 1–2 mm long. Flowers 10–16 in a dense subglobose head, white or dark wine-red; pedicel and ovary 3–5 mm long. Dorsal sepal ovate, 3–4 × 2 mm. Lateral sepals concave, elliptic, obtuse, 3–4 × 2–2.5 mm. Petals spathulate, 1–1.5 × 0.7–0.8 mm. Lip arcuate, fleshy, circular, 1.5 mm in diameter, papillose in front; callus of 2 converging ridges in basal half of lip. Column c. 1 mm long.

DISTRIBUTION: Tutuila and 'Upolu. Also found in New Caledonia and Vanuatu.

HABITAT: Uncommon epiphyte in lowland to montane forest; 225–700 m.

COLLECTIONS: *Christophersen* 1185 (BISH); *Whistler* 712 (HAW), 712a (HAW), 744 (HAW), 2050 (HAW), 2902 (HAW), 6999 (HAW).

section BRACHYPUS

B. samoanum *Schltr.* in Repert. Spec. Nov. Regni Veg. 9: 107 (1910). Type: Samoa, *Vaupel* 546 (holo. B†).

B. sp. 2 sensu Christophersen in Bernice P. Bishop Mus. Bull. 128: 68 (1935).

B. christophersenii L.O.Williams in Bot. Mus. Leafl. 7: 143 (1939). Type: Samoa, *Christophersen* 2297 (holo. AMES!, iso. BISH!, K!, P!).

An epiphyte with a clustered or weakly spreading habit; rhizome short, c. 2 mm in diameter. Pseudobulbs pear-shaped, 1.2–1.5 × 0.7–0.8 cm. Leaves oblong-oblanceolate, obtuse, 6.5–14 cm long, 1–2.2 cm wide, slenderly petiolate. Inflorescence erect, one-flowered, 6–12 cm long; peduncle short, terete, 1–3.5 cm long; bract minute, ovate, apiculate. Flower greenish yellow with reddish purple markings; pedicel and ovary 5–8.5 mm long. Dorsal sepal lanceolate, acute, 11–12 × 3–5 mm. Lateral sepals obliquely oblong to oblong-elliptic, acuminate, 13.5–15 × 4–5.5 mm. Petals linear-ligulate, obliquely subacute or obtuse at apex, 2.5–4 × 1.5–2 mm. Lip very fleshy, slightly arcuate, ovate-sublinguiform, auriculate at base, obtuse at apex, 2.5–4 × 1.5–2 mm; disc sulcate, glabrous. Column 3–4 mm long with two subulate apical stelidia; foot 2 mm long.

DISTRIBUTION: Ofu, Olosega, Savai'i, Ta'u, Tutuila and 'Upolu. Also in Vanuatu, New Caledonia and Fiji.

HABITAT: Occasional epiphyte in montane and cloud forest; 300–800 m.

COLLECTIONS: *Betche* s.n. (MEL); *Christophersen* 277 (BISH), 2246 (BISH, K), 2258 (BISH), 2297 (AMES, BISH, K, P); *Vaupel* 546 (B†); *Whistler* 23 (BISH, HAW), 591 (HAW), 901 (HAW, K), 1562 (HAW), 2787 (HAW, K), 3251 (HAW, K), 3736 (HAW, K), 5131 (HAW), 5148 (HAW), 7058 (HAW), 7984 (HAW), 8150 (HAW), 8295 (HAW), 8496 (HAW), 8813 (HAW); *Yuncker* 9153 (BISH), 9249 (BISH).

section MACROURIS

B. savaiense *Schltr.* in Repert. Spec. Nov. Regni Veg. 9: 106 (1910). Type: Samoa, *Vaupel* 596 (holo. B†).
B. sp. 1 sensu Yuncker in Bernice P. Bishop Mus. Bull. 184: 33 (1935).

Tiny hanging or erect epiphytes, 2.5–6 cm long, with a short rhizome. Pseudobulbs clustered, narrowly ovoid, 4–11 × 2–3.5 mm. Leaves elliptic to oblong-elliptic, obliquely acute and minutely apiculate at apex, 10–25 × 3–4.5 mm, sessile. Inflorescence solitary, erect, racemose, 2–5.5 cm tall, laxly 3–6-flowered; peduncle filiform, a third to a half as long as inflorescence; rachis weakly fractiflex; bracts transversely ovate, acuminate, 1–1.5 mm long. Flowers minute, white, glabrous; pedicel and ovary c. 1 mm long. Dorsal sepal ovate-attenuate, acute, 0.9 × 0.3 mm. Lateral sepals obliquely elliptic-ovate, acute, 1 × 0.3 mm. Petals oblanceolate, rounded at apex, 0.5 × 0.2 mm. Lip slightly fleshy, ovate-subrhombic, obtuse, 0.4 mm long and wide, lacking a callus. Column 0.3 mm long; foot 0.2 mm long.

DISTRIBUTION: Olosega, Savai'i, Ta'u and Upolu. Also from Vanuatu and Fiji.

HABITAT: Uncommon epiphyte in montane forest; 300 to 600 m.

COLLECTIONS: *Christophersen* 2244 (BISH), 2301 (BISH); *Vaupel* 596 (B†); *Whistler* 25 (HAW), 584 (HAW), 3735 (BISH, HAW), 3816 (HAW), 4702 (HAW), 5153 (HAW), 6933 (HAW); *Yuncker* 9272 (BISH).

section MICROMONANTHE

B. distichobulbum *Cribb* in Kew Bull. 50: 787 (1995). Type: Samoa, Tutuila, *Whistler* 3757 (holo. BISH!).

A small epiphyte with a short ascending rhizome. Pseudobulbs distichously arranged, proximate, subspherical, angular, 4–5 mm in diameter. Leaves coriaceous, elliptic or elliptic-obovate, obtuse to subacute, 2.1–4 × 1–1.5 cm. Inflorescence much longer than the leaf, one-flowered, glabrous; peduncle slender, filamentous, 5–5.5 cm long; bract sheathing, acute to acuminate, 4 mm long. Flower yellow; pedicel and ovary 5–5.5 cm long. Dorsal sepal linear-lanceolate, acuminate, 11–14 × 1.5–2.1 mm. Lateral sepals obliquely lanceolate, acuminate, 14–14.5 × 2–2.5 mm. Petals transversely oblong, aristate at apex, 3 × 1 mm. Lip fleshy, narrowly ellipsoidal, auriculate at base, rounded in front, 4.5 × 1.5 mm, hairy; basal auricles erect; callus of two short ridges between auricles. Column 2.5 mm long including 1.5 mm long apical aristate stelidia; foot swollen at base, 1.5 mm long.

DISTRIBUTION: Tutuila. Endemic.
HABITAT: Rare epiphyte on tree trunks in ridge-top forest; 130–270 m.
COLLECTIONS: *Whistler* 3757 (BISH, HAW), 8714 (HAW), 9348 (HAW, K).

NOTE: Closely related to the Fijian endemic species *B. aristopetalum* Kores but having larger flowers and a longer inflorescence overtopping the leaf.

section POLYBLEPHARON

B. membranaceum *Teijsm. & Binnend.* in Natuurk. Tijdschr. Ned. Ind. 3: 397 (1855). Type: Java, *Teijsman & Binnendijk* s.n. (holo. BO!).
B. prenticei sensu Kraenzl. in Bot. Jahrb. Syst. 25: 607 (1898), non F.Muell.
B. nuruanum Schltr. in K. Schum. & Lauterb., Nachtr. Fl. Deutsch. Sudsee: 212 (1905). Type: New Guinea, *Schlechter* 13800 (holo. B†).
B. betchei [as *Bolbophyllum betschei*] sensu H.Fleischm. & Rech. in Denkschr. Kaiserl. Akad. Wiss., Math.-Naturwiss. Kl. 85: 261 (1910), non F.Muell.
B. gibbonianum Schltr. in Bot. Jahrb. Syst. 56: 483 (1921). Type: Palau, *Ledermann* 14549 (holo. B†, iso. K!).

A creeping epiphyte with an elongate rhizome. Pseudobulbs borne 2.5–6 cm apart, ovoid, 6–7 × 5 mm, unifoliate at apex. Leaves ovate, acute, 2–4 × (0.7–) 1–1.5 cm. Inflorescences arising from the base of the pseudobulbs or along the rhizome, 2 mm long, bearing a solitary flower; bract elliptic, 1–1.5

Fig. 24. *Bulbophyllum distichobulbum.* **A**, habit × 1; **B**, flower × 3½; **C**, flower, dissected view × 3½; **D**, column and lip, side view × ⅔; **E**, lip × 10; **F**, column, dorsal view × 10; **G**, column, side views × 17; **H**, column, ventral view × 17; **J**, fruiting inflorescence × 1. All drawn from the type collection by Susanna Stuart-Smith.

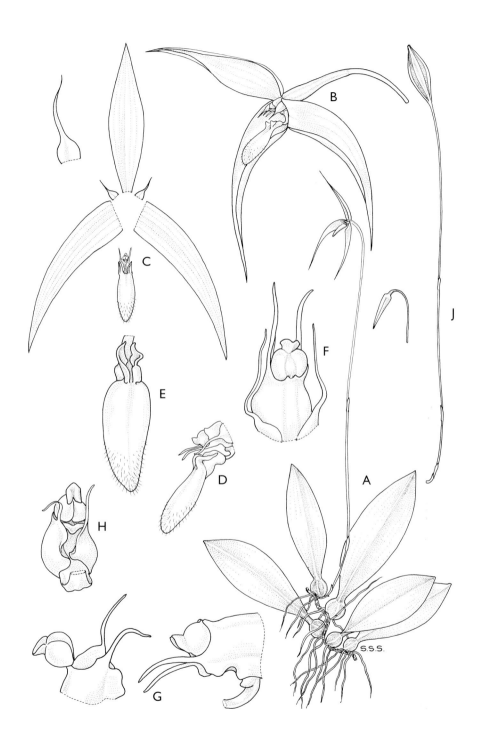

mm long. Flower fleshy, not opening widely, pale translucent yellow with a purple lip; pedicel and ovary 5–7 mm long, ovary 6-angled. Dorsal sepal oblong-ovate, acute, 3.5–4.5 × 1.5 mm; lateral sepals connate, obliquely ovate, acute, 4 × 2–2.5 mm. Petals subcircular-obovate, apiculate, 1.5 × 0.8–1.2 mm. Lip recurved, oblong to ligulate, 1.5–2 × 0.8 mm, papillose; callus of 2 longitudinal ridges. Column 1 mm long; stelidia truncate, bidentate; foot 1.2–1.5 mm long.

DISTRIBUTION: Savai'i, Tutuila and 'Upolu. Also found in Thailand, Malay Peninsula, Malay Archipelago, New Guinea, Solomon Islands, Vanuatu, Fiji and Tonga.

HABITAT: Occasional climbing epiphyte in mangrove to montane forest; sea level to 820 m.

COLLECTIONS: *Betche* s.n. (MEL); *Christophersen* 12 (BISH), 47 (BISH); *Powell* 147 (K), s.n. (K); *Rechinger* 592 (BM, K, W), 1682 (W), 1842 (W); *Reinecke* 42 (B†), 42a (B†); *Vaupel,* 412 (K); *Whistler* 440 (HAW, K), 799 (HAW), 1572 (HAW), 2901 (BISH, HAW), 4142 (BISH, HAW), 6858 (HAW), 7013 (HAW), 8259 (HAW), 8695 (HAW); s.n. (BM); *Whitmee* 263 (K), s.n. (BM); *Wisner* 140 (BISH).

section CIRRHOPETALUM

B. longiflorum *Thouars,* Orch. Iles. Austr. Afr.: 98 (1822). Type: Mauritius, *Thouars* s.n. (holo. P!).

Epidendrum umbellatum G.Forst. in Ins. Austr. Prodr.: 60 (1786). Type: Society Islands, *G.Forster* s.n. (holo. BM!).

Cymbidium umbellatum (G.Forst.)Sprengel in Pl. Min. Cog. Pugill. 2: 82 (1815).

Cirrhopetalum thouarsii Lindl. in Bot. Reg. 10: 832 (1824). Type: Fiji, *Seemann* 598 (holo. K!).

C. umbellatum (G.Forst.)Hook. & Arn. in Bot. Beechey Voy.: 71 (1832).

C. clavigerum Fitzg. in J. Bot. 21: 204 (1883). Type: N. Australia, *Fitzgerald* s.n. (holo. BM!).

Bulbophyllum clavigerum (Fitzg.)F.Muell. in Syst. Cens. Austr. Pl. Suppl. 1: 3 (1884).

Phyllorchis umbellata (G.Forst.)Kuntze in Rev. Gen. Pl. 2: 657 (1891).

P. thouarsii (Lindl.)Kuntze, l.c. 677.

Cirrhopetalum thouarsii var. *concolor* Rolfe in Gard. Chron. ser. 3, 12: 178 (1892). Type: Not found.

C. kenejianum Schltr. in Repert. Spec. Nov. Regni Veg. Beih. 1: 889 (1913). Type: New Guinea, *Schlechter* 18462 (holo. B†).

C. longiflorum (Thouars) Schltr. in Beih. Bot. Centralbl. 33, 3: 420 (1915).

A creeping epiphyte with an elongated rhizome, 3–5 mm in diameter. Pseudobulbs widely spaced, 2.5–7.5 cm apart, obliquely ovoid, 1.7–3.5 × 0.8–1.5 cm, 4-angled. Leaves oblong-lanceolate to oblong-elliptic, obtuse, 6–19 × 1.5–4 cm, slenderly petiolate. Inflorescences erect, umbellate, 15–30 cm tall, 3–7-flowered; peduncle slender, wiry; bracts lanceolate, acuminate, 6–10 mm long. Flowers large, dull greenish yellow to pale yellow, with or without copious purple spotting on sepals and petals; pedicel and ovary 1.2–2 cm long. Dorsal sepal elliptic-ovate, with a slender filiform apex, 5–8 × 4–5 mm. Lateral sepals connate along upper margins, ligulate, acute, 25–44 × 3–5 mm. Petals elliptic-oblong, filiform-attenuate at apex, 3–5 × 1–2.5 mm, laxly long-ciliate on upper margins. Lip fleshy, arcuate, sublinguiform, attenuate, weakly auriculate at base, obtuse, 4.5–5.5 × 1.5 mm, with 2 prominent raised ridges on basal half. Column 3 mm long with 2 apical subulate stelidia; foot c. 3.5 mm long.

DISTRIBUTION: Savai'i and 'Upolu. Widespread from Africa, and Madagascar to the Malay Archipelago, Australia, New Guinea, New Caledonia, Fiji and eastwards to the Society and Austral Islands.

HABITAT: Rare epiphyte in montane forest; 300 m.

COLLECTIONS: *Flynn* 3600 (NTBG); *Whistler* 1769 (HAW), 1987 (HAW).

40. **THRIXSPERMUM**

Lour., Fl. Cochin.: 519 (1790)

Erect to pendulous epiphytic or rarely lithophytic or terrestrial herbs. Stems short to long, leafy, unbranched. Leaves coriaceous to fleshy, few to many, distichous, articulated to tubular sheaths. Inflorescences lateral, few- to many-flowered, simply racemose; rachis with spiralling or distichous bracts. Flowers biseriate or all around stalk, resupinate, small to relatively large, sequentially produced, short-lived. Sepals and petals free, spreading, the lateral sepals decurrent on column-foot. Lip adnate to column-foot, immobile, concave, 3-lobed, with a callus; side lobes erect; midlobe short to long, fleshy. Column short, stout, with a foot; pollinia 2, waxy, attached by a short broad stipe to a small viscidium.

A genus of some 120 species distributed in tropical Asia, S.E. Asia, Malesia, New Guinea, the Philippines, Micronesia, N.E. Australia, the Solomon Islands, Vanuatu, New Caledonia, Fiji and Samoa. A single species in Samoa.

T. graeffei *Rchb.f.* in Seem., Fl. Vit.: 297 (1868). Type: Samoa, Upolu, *Graeffe* s.n.(holo. W!, iso. BM!, record AMES!).

Sarcochilus graeffei (Rchb.f.)Benth. & Hook.f. ex Drake, Ill. Ins. Mar. Pacif.: 510 (1892).

Sarcochilus sp. nov. sensu Kraenzl. in Bot. Jahrb. Syst. 25: 608 (1898).

Thrixspermum oreadum Schltr. in Repert. Spec. Nov. Regni Veg. Beih. 1: 962 (1913). Type: New Guinea, *Schlechter* 16927 (holo. B†).

T. sp. sensu Yuncker in Bernice P. Bishop Mus. Bull. 184: 33 (1945).

Small pendent epiphytic plants, 6–12 cm long. Stems short, 3–4 mm in diameter, obscured by leaf bases. Leaves closely spaced, ligulate to ligulate-lanceolate, obliquely acute or subacute, 5.5–11 × 0.8–1.3 cm, articulated to a 0.5–0.9 cm long sheathing leaf-base. Inflorescences erect or ascending, 7–14 cm long, many-flowered; rachis half to quarter length of inflorescence; bracts slightly equitant, 1.5–2 mm long. Flowers white, short-lived. Dorsal sepal oblong-elliptic, obtuse, 5–5.5 × 2.5–3 mm. Lateral sepals slightly oblique, 5–5.5 × 2.5–3 mm. Petals elliptic, obtuse, 4.5–5.2 × 2.5 mm. Lip weakly 3-lobed, subcordate, 3.5–4 × 4.5 mm, bilaterally somewhat compressed, minutely pubescent on disc; side lobes upcurved, broadly rounded; midlobe small, subquadrate to broadly spathulate. Column 1 mm long; foot 0.5 mm long. Fruit cylindrical, 4–5 cm long.

DISTRIBUTION: Savai'i, Ta'u, Tutuila and 'Upolu. Also in New Guinea, the Solomon Islands, Vanuatu and Fiji.

HABITAT: Epiphyte in montane forest; 400–725 m.

COLLECTIONS: *Christophersen* 1180 (BISH, K), 1192 (BISH), 3252 (BISH); *Graeffe* s.n. (BM, W); *Huegel* 513 (K); *Powell* 357 (BISH, K); *Rechinger* 95 (W), 131 (W), 195 (W), 1684 (W), 1700 (W); *Reinecke* 237 (B†); *Vaupel* 151 (AMES, BM, K); *Walter* s.n. (BISH); *Whistler* 2041 (HAW), 3749 (HAW), 3852 (HAW), 3879 (HAW, K), 5152 (HAW), 5719 (HAW), 6870 (HAW); *Whitmee* 44 (K), s.n. (K); *Yuncker* 9271 (BISH).

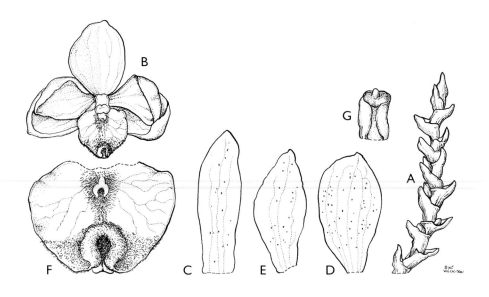

Fig. 25. *Thrixspermum graeffei*. **A**, rachis and bracts × 4; **B**, flower × 4; **C**, dorsal sepal × 8; **D**, lateral sepal × 8; **E**, petal × 8; **F**, lip × 10; **G**, column × 14. All drawn from *Nicholls* 5 by Sue Wickison.

41. **SCHOENORCHIS**

Reinw. ex Blume, Cat. Buitenz.: 100 (1823) nom.nud. & Bijdr. Fl. Ned. Ind.:
361 (1825); Reinw. in Hornsch., Syll. Pl. Nov. 2: 4 (1825)

Erect to pendulous small to medium-sized epiphytic herbs. Stems short to
elongate, usually branched, leafy. Leaves distichous, coriaceous or fleshy,
linear or terete, articulated to a sheathing tubular leaf-base. Inflorescences
lateral, erect to pendent, racemose or paniculate, densely to laxly few- to
many-flowered. Flowers resupinate, small. Sepals and petals free, connivent
or scarcely spreading. Lip fleshy, immobile, 3-lobed, with a basal spur more or
less perpendicular to lamina; side lobes erect; midlobe laterally compressed;
spur sometimes septate. Column short, lacking a foot; pollinia 2, waxy, deeply
cleft, attached by a short stipe to a relatively large viscidium; rostellum
prominent, bilobed.

A small genus of about 20 species in tropical Asia, S.E. Asia, Malesia, the
Philippines, New Guinea, N.E. Australia, the Solomon Islands, Vanuatu, New
Caledonia, Fiji and Samoa. A single species in Samoa.

S. micrantha *Reinw. ex Blume*, Bijdr. Fl. Ned. Ind. 1(8): 362 (1825). Type: Java,
Blume s.n.(holo. L!).
Saccolabium chionanthum Lindl. in J. Linn. Soc. 3: 35 (1859). Type: Java,
   Reinwardt s.n. (holo. K!)
S. plebejum J.J.Sm. in Bull. Jard. Bot. Buitenz. ser. 2, 3: 77 (1912). Type: New
   Guinea, *Gjellerup* 21 (holo. BO!).
Schoenorchis densiflora Schltr. in Repert. Spec. Nov. Regni Veg. Beih. 1: 986
   (1914). Type: New Guinea, *Schlechter* 17316 (holo. B†, iso. K!).
Ascocentrum micranthum (Reinw. ex Blume) Holtt. in Gard. Bull. Sing. 11: 275
   (1947).

An epiphyte with clustered branching stems up to 15 cm long. Leaves fleshy,
terete, 1.5–4 × 0.2 cm, slightly channelled on dorsal surface, articulated to
0.3–0.5 cm long sheaths. Inflorescences simple, 2–5 cm long, densely many-
flowered; peduncle short; bracts c. 1 mm long. Flowers tiny, white flushed with
pink, turning yellow with age; ovary 2–3 mm long. Sepals oblong to oblong-
ovate, subacute, 1.5–2 × 0.6–0.8 mm. Petals oblique, oblong-obovate, subacute
or retuse, 1.3–1.8 × 0.3–0.5 mm. Lip closely appressed to column, 1.3–1.5 mm
long, with a low callus in mouth of spur; side lobes erect, fleshy, broader than
long; midlobe porrect, fleshy, laterally compressed; spur saccate 0.8–1 mm
long. Column 0.6 mm long. Fruit 0.8–1.2 cm long.

DISTRIBUTION: Savai'i and 'Upolu. Also widely distributed from S.E.
Asia and the Malay Archipelago to New Guinea, Bougainville, the Solomon
Islands, Vanuatu, New Caledonia, Fiji and Australia.
   HABITAT: Rare epiphyte found in lowland to montane forests; 50–1600 m.
COLLECTIONS: *Betche* s.n. (MEL); *Cox* 286 (BISH); *Whistler* 926 (HAW).

42. **LUISIA**

Gaudich. in Freycinet, Voy. Uranie et Physicienne, Bot.: t.37 (1827), op.cit.
426 (1829)

Medium-sized epiphytic herbs with long unbranched laxly leafy stems, rooting in basal part. Stems covered by tubular persistent leaf-bases, terete. Leaves terete, dull green. Inflorescences lateral, short, densely few- to several-flowered. Flowers very fleshy, small to medium-sized, resupinate, long-lasting; ovary triangular. Sepals free, spreading, narrow. Petals smaller but similar to dorsal sepal. Lip immobile, bipartite, lacking a spur; basal part narrower than apical part; apical part fleshy, entire, often longitudinally furrowed. Column short, lacking a foot; pollinia 2, waxy, entire or slightly cleft, attached by a stipe to a large viscidium; rostellum short, broad, entire or slightly bifid.

A genus of about 36 species in tropical Asia, S.E. Asia, Malesia, New Guinea, N.E. Australia, the Marianas and the S.W. Pacific Islands across to New Caledonia, Samoa and Fiji. A single species reported from Samoa.

L. teretifolia *Gaudich.* in Freycinet, Voy. Uranie et Physicienne, Bot.: t.37 (1827), op. cit.: 426(1829). Type: Mariana Islands, Guam, *Gaudichaud* 37 (holo. P!). *L. sp.* sensu Schltr. in Repert. Spec. Nov. Regni Veg. 9: 110 (1911).

An erect epiphyte 15–30 cm long. Stems rigid, up to 4 mm in diameter. Leaves suberect, terete, longitudinally striate, 4–20 × 0.3 cm. Inflorescences few-flowered, 0.4–1 cm long; bracts 1–2 mm long. Flowers with green or yellowish green sepals and petals and a reddish brown to dark maroon lip; ovary 3–5 mm long. Dorsal sepal elliptic, subacute to obtuse, 5–5.5 × 3 mm. Lateral sepals lanceolate, fleshy at apex, 7–8 × 2.5 mm, slightly keeled on back. Petals linear-oblong, obtuse, 3–7 × 3.5–4 mm. Lip bipartite, 6.5–7.5 cm long; basal part subquadrate, with small elliptic-falcate side lobes; apical part broadly cordate, obtuse. Column 2 mm long.

DISTRIBUTION: Savai'i only. Also widely distributed from tropical Asia to the Malay Archipelago, New Guinea, the Mariana Islands (Guam), the Solomon Islands, Vanuatu, New Caledonia and Australia.

HABITAT: Rare epiphyte found in rain-forest and on scrub in lava fields; 50 – 500 m..

COLLECTIONS: *Vaupel* 650 (B†); *Whistler* 948 (HAW), 8271 (HAW).

43. **MICROTATORCHIS**

Schltr. in K.Schum. & Lauterb., Nachtr. Fl. Deutsch. Schutzgeb. Sudsee:
224 (1905).

Tiny epiphytic herbs with more or less flattened photosynthetic ribbon-like roots. Stems very short, often lacking leaves for lengthy periods. Leaves clustered, small, oblong-obovate, membranous, articulate, deciduous. Inflorescences lateral, erect, racemose, many-flowered; rachis more or less fractiflex, often angled or winged; bracts persistent, small, often with tiny stipule-like appendages at base. Flowers sequential, very small, membranous, pale green or yellow, resupinate. Sepals and petals fused in basal half to form a tubular perianth, free in distal parts. Lip immobile, entire or weakly 3-lobed, apex often retrorsely inflexed, shortly spurred at base; spur scrotiform or saccate, retrorse, appendaged within. Column short, with two apical short stelidia, lacking a column-foot; pollinia 2, waxy, subglobose, entire, attached by a stipe to a small viscidium.

A small genus of about 45–50 species in Malesia across to the Pacific Islands eastwards to the Society Islands and south to New Caledonia. Centre of diversity in New Guinea. A single species reported from Samoa.

M. samoensis *Schltr.* in Repert. Spec. Nov. Regni Veg. 9: 111 (1910). Type: Samoa, *Vaupel* 470 (holo. B†).

A dwarf epiphyte 2–4 cm tall. Roots filiform, up to 1 mm in diameter. Leaves oblanceolate, acute, 0.8–1.5 × 2–3 mm. Inflorescences erect, 2–3 cm long, laxly few-flowered; peduncle more or less equalling rachis; bracts small, scale-like, ovate, acute, 0.5–0.8 mm long. Flowers greenish yellow. Sepals and petals lanceolate, acuminate, fused in lower third, 1.5–2 mm long. Lip narrowly elliptic-ovate, 1.5–2 × 0.4–0.6 mm, the apex with a small appendage, 0.2–0.3 mm long; side lobes somewhat incurved; spur globose, c. 0.5 mm long.

DISTRIBUTION: Savai'i and 'Upolu. Also from Fiji.
HABITAT: Epiphyte found in cloud-forest; 800–900 m.
COLLECTIONS: *Christophersen* 2268 (BISH); *Vaupel* 470 (B†); *Whistler* 7083 (BISH, HAW).

44. **POMATOCALPA**

Breda, Gen. Spec. Orchid. Asclep.: 29, t.15 (1829)

Epiphytic herbs with short to long erect leafy stems. Leaves few to many, oblong to ligulate, fleshy or coriaceous, articulated at base to imbricate sheathing bases. Inflorescences lateral, erect or pendent, simple or branching, usually densely many-flowered. Flowers small, non-resupinate, fleshy. Sepals and petals free, spreading, subsimilar. Lip immobile, 3-lobed, spurred at base; spur inflated distally or not, with a prominent ligule on back wall extending into mouth of spur; midlobe fleshy. Column short, stout, lacking a foot; rostellum small, more or less malleiform in shape; anther terminal, operculate; pollinia two, subglobose, attached by a linear stipe to a small viscidium.

A genus of some 35 species in tropical Asia, the Malay Archipelago, S.W. Pacific Islands, Micronesia and N.E. Australia. A single species reaches Samoa.

P. vaupelii *(Schltr.) J.J.Sm.* in Natuurk. Tijdschr. Ned.-Indie 72: 107 (1912). Type: Samoa, *Vaupel* 323 (holo. B†, iso. K!).

Cleisostoma spathulata sensu H.Fleischm. & Rech. in Denkschr. Kaiserl. Akad. Wiss., Math.-Naturwiss. Kl. 85: 261 (1910), non Blume.

Saccolobium vaupelii Schltr. in Repert. Spec. Nov. Regni Veg. 9: 110 (1910).

S. spec. aff. *ramuloso* Lindl.; Schltr. in Repert. Spec. Nov. Regni Veg. 9: 110 (1910).

Robiquetia spathulata auct. non (Blume) J.J.Sm.

An erect or pendent herb. Stem up to 30 cm long, c. 8 mm in diameter, completely covered by sheathing leaf bases. Leaves elliptic to ligulate, unequally roundly bilobed at apex, 6–20 × 1.8–3.8 cm; sheaths 1.2–2.2 cm long, prominently striate. Inflorescences pendent, racemose or weakly branching, 10–26 cm long; branches 2–5.5 cm long; peduncle short, terete, 1.5–3 cm long; bracts reflexed, broadly deltoid, 1–3 mm long. Flowers fleshy, yellow with dark red or brownish markings towards the base of the segments, lepidote on outer surface; pedicel and ovary 2–4 mm long, lepidote. Dorsal sepal oblong or oblong-obovate, obtuse, 4–7 × 2–4 mm. Lateral sepals obliquely obovate, broadly acute, 4.5–6 × 2.5–4.5 mm. Petals narrowly oblong-obovate, rounded at apex, 3.8–6 × 2–3 mm. Lip porrect, 3-lobed, 2–3 mm long; side lobes erect, broadly obtuse, 0.8–1.2 × 1.5–1.8 mm, somewhat thickened basally; midlobe small, fleshy, transversely ovate, truncate, 2–2.5 × 3–3.5 mm; spur saccate, 2–4 × 2–3 mm, with a prominent ascending appendage on back wall almost blocking mouth of spur. Column 1.5–2 mm long.

DISTRIBUTION: Savai'i and Upolu. Also found in Fiji.

HABITAT: Epiphyte found in montane forest; 400–600 m.

COLLECTIONS: *Rechinger* 1673 (W); *Vaupel* 323 (K), 652 (B†); *Whistler* 161 (HAW), 5367 (BISH, HAW), 9403 (HAW, K), 9503 (HAW); *Whitmee* s.n. (K).

45. TAENIOPHYLLUM

Blume, Bijdr. Fl. Ned. Ind.: 355 (1825)

Small to tiny acaulescent epiphytic herbs with terete to stongly dorso-ventrally flattened photosynthetic roots. Stem very short, bearing scales. Inflorescences erect, short to long, unbranched; peduncle and rachis glabrous or hairy; rachis often fractiflex. Flowers small, distichous, ephemeral or lasting a few days, pale green, yellow or white. Sepals and petals free or fused in basal part to form a distinct perianth tube, apical parts free. Lip immobile, entire or 3-lobed, simple or with a retrorse bristle-like apex, spurred at base, the mouth often covered partially or entirely by a hyaline septum. Column short, lacking a foot; pollinia 4, waxy, obovoid to ellipsoidal, attached by a slender stipe to a large viscidium.

A large genus of about 170 species in subtropical and tropical Asia and S.E. Asia, Malesia, New Guinea, the Philippines, Micronesia, N.E. Australia and the

S.W. Pacific Islands across to the Austral Islands and Pitcairn and south to New Caledonia. One species in tropical Africa. Three species reported from Samoa.

1. Inflorescence short, less than 2 cm long, verruculose on peduncle and rachis · **T. fasciola**
   Inflorescence elongate, 3–8 cm long, glabrous · · · · · · · · · · · · · · · · **2**
2. Spur conical to subcylindrical, 1.5 mm long; bracts c. 1 mm long
   · **T. whistleri**
   Spur scrotiform, c. 1 mm long; bracts c. 0.5 mm long · · · · **T. savaiiense**

T. fasciola *(G.Forst.)Rchb.f.* in Seem., Fl. Vit.: 296 (1868).Type: Tahiti, *G.Forster* 172 (holo. BM!, iso. P!).
Epidendrum fasciola G.Forst., Fl. Ins. Austr. Prodr.: 60, n.320 (1786).
Taeniophyllum seemannii Rchb.f. in Seem., Fl. Vit.: 296 (1868). Type: Fiji, *Seemann* 593 (holo. W!, iso. L!).
T. asperulum Rchb.f., Otia Bot. Hamburg. 1: 53 (1878). Type: Society Islands, *Wilkes* s.n. (holo. W!).
T. decipiens Schltr. in Repert. Sp. Nov. Regni Veg. 9: 112 (1910). Type: Samoa, *Vaupel* 278 (holo. B†, iso. AMES!, B!, BISH!, K!, W!).
T. parhamiae L.O.Williams in Bot. Mus. Leafl. 7: 148 (1939). Type: Fiji, *Parham* 3 (holo. AMES!).
T. fasciola var. *mutina* N. Hallé in Fl. Nouv.-Caled. Dépend. 8: 386, pl.157 (1977). Type: New Caledonia, *McKee* 29285 (holo. P!).

Small epiphytic plant with long flattened spreading roots 10–30 × 2–5 mm. Inflorescences short, densely few-flowered; peduncle and rachis densely verruculose; rachis fractiflex, 1–4 cm long; bracts ovate, 1 mm long. Flowers pale yellowish white, with darker yellow tips to the perianth segments; ovary glandular-pubescent. Sepals free, ovate- to ovate-elliptic, subacute, 2.5–4 × 1–2 mm. Petals free, ligulate or elliptic-ligulate, subacute, 2.5–4 × 0.5–1 mm. Lip entire, concave, navicular, 1.5–2.2 mm long; spur perpendicular to lip, short, subcylindrical to conical, subacute to obtuse, 1–2 mm long. Column very short; pollinia 4, ellipsoidal to obovoid.

DISTRIBUTION: Aunu'u, Ofu, Olosega, Savai'i, Ta'u, Tutuila and 'Upolu. Also in the Mariana Islands (Guam), the Solomon Islands, Vanuatu, New Caledonia, Fiji, Tonga, the Cook Islands, the Horne Islands, Tahiti, the Austral Islands and Pitcairn Island.
HABITAT: Epiphyte found in coastal to lower montane forests; sea level to 300 m.
COLLECTIONS: *Garber* 641 in part (BISH); *Rechinger* 200 (BM, K), 1892 (W), 5304 (W); *Vaupel* 278 (AMES, B, BISH, K, W); *Whistler* 1587 (HAW), 1877 (HAW), 2918 (HAW), 2984 (BISH, HAW), 3004 (BISH, HAW, K), 3073 (HAW), 3117a (HAW), 3343 (HAW), 3765 (HAW), 3787 (HAW), 4468 (HAW, K), 8657 (HAW), 9101 (HAW); *Whitmee* 31 (K); *Wilkes* s.n. (W); without collector (BM, K).

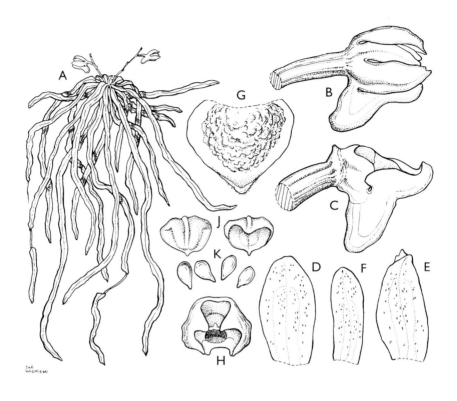

Fig. 26. *Taeniophyllum fasciola*. **A**, habit × ⅔; **B**, flower × 6; **C**, column, lip and ovary × 6; **D**, dorsal sepal × 8; **E**, lateral sepal × 8; **F**, petal × 8; **G**, lip × 10; **H**, column apex × 10; **J**, anther cap two views × 8; **K**, pollinia × 8. All drawn from *Cribb & Morrison* 1781 by Sue Wickison.

T. whistleri *Cribb* **sp. nov.** affinis *T. smithii* Kores sed radicibus regionibus pneumaticis deficientibus, inflorentiis crassioribus glabris, labello integro ovato, calcari ellipsoideo angulato differt. Typus: Samoa, Tutuila, *Whistler* 3117 (holotypus HAW!, isotypus K!).

A tiny epiphyte with roots appressed to substrate or free and pendulous, 3.5–20 × 0.2–0.4 cm. Inflorescences 1–few, laxly few-flowered, unbranched, 2.5–8 cm long; peduncle filiform, glabrous; rachis fractiflex, 0.5–1.6 cm long; bracts ovate, acute, 0.8–1 mm long. Flowers greenish yellow to creamy white. Sepals and petals free, erect or weakly spreading, lanceolate to oblong-lanceolate, acute or subacute, 2.8–3.7 × 0.7–1.2 mm. Lip ovate, obscurely 3-lobed, 2.5–3 × 2–2.5 mm; side lobes upcurved, obtuse; midlobe fleshy, subacute; spur retrorse, incurved, subsaccate-cylindric, c. 1.5 mm long. Column short, broad.

DISTRIBUTION: Savai'i, Ta'u and Tutuila. Endemic.

HABITAT: Uncommon epiphyte found in lowland and montane forests; sea leavel to 580 m.

COLLECTIONS: *Garber* 641 in part (BISH); *Whistler* 3117 (HAW, K); *Yuncker* 9309a (AMES, BISH).

NOTE: Close to *T. smithii* Kores from Fiji but it seems to be distinct; the inflorescences are stouter, the roots are more undulate and lack pellucid pneumatothodes, the lip is ovate rather than 3-lobed and the spur is incurved and somewhat 4-angled in cross-section. However, the specimens lack good flowers and the measurements given here refer to withered flowers retained at the apex of developing fruits. It is also close to the Fijian *T. gracile* (Rolfe) Garay but with narrower roots lacking the prominent surface marking of the type. Kores (1991) included *T. vitiense* L.O.Williams as a synonym of *T. gracile*, but the former has a glandular inflorescence axis while the latter is glabrous.

T. savaiiense *Cribb et Whistler* **sp. nov.** affinis *T. elegantissimo* Rchb.f. habitu similis sed labello obscure trilobato, calcari scrotiformi vel mamilliformi distinguenda; a *T. kompsopo* Schltr. radicibus latioribus, bracteis brevioribus distichis, calcari pendenti sessili distinguenda. Typus: Samoa, Savai'i, *Whistler* 970 (holotypus HAW!).

A tiny epiphyte with many narrow ribbon-like roots up to 11 × 1–2 mm. Inflorescences 1.5–5 cm long; peduncle filamentous, wiry, glabrous, up to 3 mm long; rachis glabrous, appearing fractiflex; bracts distichous, 0.5 mm apart, triangular, acute, 0.5 mm long. Flowers minute, white or cream-coloured; pedicel and ovary 0.5 mm long. Dorsal sepal oblong, acute, 1.5–2 × 0.7 mm. Lateral sepals obliquely oblong-lanceolate, acute, 1.5–2 × 1 mm. Petals oblong, obtuse, 2 × 0.5 mm. Lip obscurely 3-lobed, ovate, acute but fleshy at apex, 1.5 × 0. 7 mm; spur scrotiform, 1 mm across.

DISTRIBUTION: Savai'i only. Endemic.
HABITAT: Epiphytic on tree trunks in montane forest; 15–440 m.
COLLECTIONS: *Whistler* 26 (BISH, HAW), 927 (HAW), 970 (HAW), 5146 (HAW).

NOTE: Close to *T. elegantissimum* Rchb.f. but the spur is scrotiform with an apical nipple rather than inflated and cylindrical as shown in Reichenbach's drawing of the type.

46. CHRYSOGLOSSUM

Blume, Bijdr. Fl. Ned. Ind.: 337 (1825)

Terrestrial herbs with creeping rhizomes. Pseudobulbs erect, small, several-noded, unifoliate. Leaves erect, petiolate, articulate. Inflorescences terminal, elongate, racemose, laxly few- to many-flowered, borne on leafless

rudimentary pseudobulbs alternating with vegetative shoots. Flowers small to medium-sized, resupinate or not. Sepals and petals subsimilar, free, erect or spreading, lateral sepals decurrent on column-foot to form a short to long mentum. Lip adnate to column-foot, entire to 3-lobed, somewhat motile, with a callus of one or more longitudinal keels. Column fairly long, with a foot that is saccate or spur-like at base; pollinia 2, waxy, unappendaged; rostellum short.

A genus of some 25 species in tropical Asia, S.E. Asia, Malesia, New Guinea, the Philippines, the Solomon Islands, Vanuatu, New Caledonia, Fiji and Samoa. A single species reported from Samoa.

C. ornatum *Blume*, Bijdr. Fl. Ned. Ind.: 338 (1825). Type: Java, *Blume* s.n. (holo. L, iso. K!).

C. neocaledonicum Schltr. in Bot. Jahrb. Syst. 39: 58 (1906). Type: New Caledonia, *Schlechter* 15477 (holo. B†, iso. K!, P).

C. gibbsiae Rolfe in J. Linn. Soc. 39: 175 (1909). Type: Fiji, *Gibbs* 886 (holo. BM!).

A terrestrial herb 30–65 cm tall. Pseudobulbs laxly spaced on rhizome, erect, narrowly conical, 2.5–4 × 0.5–0.9 cm. Leaves erect, elliptic to oblong-elliptic, acute or acuminate at apex, 20–30 × 5.5–9.5 cm; petiole 5–11 cm long. Inflorescence erect, 30–65 cm tall, simple, laxly several-flowered; bracts 5–6 mm long. Flowers yellow or greenish yellow with red markings on the lip; ovary 10–15 mm long. Dorsal sepal oblong-lanceolate, acute, 12–17 × 3–4 mm. Lateral sepals falcate, lanceolate-oblong, acute, 10–15 × 2.5–3.5 mm; mentum obtuse, 1.5–2 mm long. Petals falcate, lanceolate, acute, 13–18 × 2.5–3 mm. Lip 3-lobed in middle, 9–11 × 5–6 mm, saccate at base, with basal margins undulate; side lobes erect, falcate-rounded; midlobe ovate, with upturned sides, obtuse; callus 3-ridged, the outer ridges reaching middle of midlobe, the central one much shorter. Column slender, 7–8 mm long, with an auricle half way along the lower margin on each side.

DISTRIBUTION: 'Upolu only. India east to S.E. Asia, Malesia, New Guinea, the Solomon Islands, Vanuatu, New Caledonia and Fiji.

HABITAT: Rare in montane forest; c. 600 m.

COLLECTION: *Whistler* 1107 (HAW).

Fig. 27. *Taeniophyllum savaiiense*. **A**, habit × 1; **B**, inflorescence × 3; **C**, flower × 15; **D**, dorsal sepal, petals and lip × 15; **E**, lip and spur × 18; **F**, lip, spur and ovary × 18; **G**, column, two views × 26. *T. whistleri*. **H**, habit × 1; **J**, inflorescence × 3; **K**, dorsal sepal, petal and lateral sepal × 10; **L**, lip and spur × 12; **M**, fruit × 3. **A–G** drawn from *Whistler* 970; **H–L** from *Whistler* 3117; **M** from *Garber* 641. All drawn by Susanna Stuart-Smith.

47. GEODORUM

Jackson in Bot. Repos. 10: t.626 (1811)

Terrestrial herbs with short subterranean rhizomes. Pseudobulbs erect, partially subterranean, one-noded, leafy at apex. Leaves few, pleated, articulated, petiolate, deciduous. Inflorescence basal, racemose, crozier-shaped at anthesis but straightening and elongating after fruit set, laxly to densely few- to many-flowered; rachis much shorter than peduncle. Flowers often showy, medium-sized, non-resupinate. Sepals and petals free, dissimilar, more or less connivent. Lip sessile, continuous with column-foot, subentire, saccate or ventricose at base, with or without a callus. Column short, broad, with a foot; pollinia 2, waxy, sulcate, attached by a short stipe to a distinct viscidium.

A genus of perhaps ten species in subtropical and tropical Asia and S.E. Asia, Malesia, New Guinea, the Philippines, Micronesia, N.E. Australia, the Solomon Islands, Vanuatu, New Caledonia, Fiji, Tonga, Niue and Samoa. A single species in Samoa.

G. densiflorum *(Lam.)Schltr.* in Repert. Spec. Nov. Regni Veg. Beih. 4: 259 (1919).Type: based on *Rheede*, Hort. Ind. Malabar. 11: 69, t. 35, 1692.
Limodorum densiflorum Lam., Encycl. Méth. Bot. 3: 516 (1792).
Cymbidium pictum R.Br., Prodr. Fl. Nov. Holl.: 331 (1810). Type: Australia, *R.Brown* 5507 (holo. BM!).
Geodorum pictum Lindl., Gen. Spec. Orch. Pl.: 175 (1833).
G. furcatum sensu Kraenzl. in Bot. Jahrb. Syst. 25: 60 (1898); non Lindl.
G. tricarinatum Schltr. in Repert. Sp. Nov. Regni Veg. 9: 101 (1910). Type: Samoa, Savai'i, *Vaupel* 285 (holo. B†), nom. nov. pro *G. furcatum* Kraenzl. non Lindl.
G. pacificum Rolfe in Bull. Misc. Inform., Kew 1908: 71 (1908). Types: Tonga, *Crosby* 246 (syn. K!); Solomon Islands, *Woodford* s.n. (syn. K!).
G. neocaledonicum Kraenzl. in Viert. Naturf. Ges. Zürich 74: 82 (1929). Type: New Caledonia, *Daeniker* 1384 (holo. Z!).

A terrestrial plant 20–30 cm tall. Pseudobulbs subglobose, clustered, 1.3–2.6 cm in diameter, covered by scarious sheaths when young. Leaves 2–4, ovate to elliptic-ovate, acute or acuminate, 9–20 × 4.5–7 cm; petiole sheath-like, 6–12 cm long. Inflorescence more or less as long as leaves; peduncle erect; rachis recurved; bracts linear-lanceolate, 1–1.3 cm long. Flowers pale pinkish white to pale purple with reddish marks and yellow blotches on lip; ovary 5–9 mm long. Sepals oblong-obovate, abruptly acuminate, 10–12 × 3–3.5 mm. Petals oblong to oblong-elliptic, obtuse to subacute, 9.5–11 × 3.5–4.5 mm. Lip cymbiform, slightly constricted in middle, weakly bilobed at apex, saccate at base, 11–13 × 10–12 mm; callus a small transverse ridge at base and warts or keels in front. Column 3 mm long; foot 3 mm long.

DISTRIBUTION: Savai'i and 'Upolu. Also found in southern China,

Burma, India, Ceylon, the southern Ryukyu Islands, New Guinea, northern Australia, the Bismark Archipelago, the Solomon Islands, Vanuatu, New Caledonia, Fiji, Tonga and Niue.

HABITAT: Rare terrestrial orchid found in fern communities and forest; altitude unknown.

COLLECTIONS: *Reinecke* 187 (B†); *Vaupel* 285 (B†).

INDEX TO SCIENTIFIC NAMES

Notes

Notes